MW01324501

topSHELF CHEMISTRY

Brian Pressley

User's Guide
to
Walch Reproducible Books

1 2 3 4 5 6 7 8 9 10

ISBN 0-8251-4625-9

J. Weston Walch, Publisher
P.O. Box 658 • Portland, Maine 04104-0658
walch.com

Printed in the United States of America

Contents

National Science Standards for High School

The goals for school science that underlie the National Science Education Standards are to educate students who are able to

- experience the richness and excitement of knowing about and understanding the natural world;
- use appropriate scientific processes and principles in making personal decisions;
- engage intelligently in public discourse and debate about matters of scientific and technological concern; and
- increase their economic productivity in their careers through using knowledge, understanding, and skills they have acquired as scientifically literate individuals.

These goals define a scientifically literate society. The standards for content define what the scientifically literate person should know, understand, and be able to do after 13 years of school science. Laboratory science is an important part of high-school science, and to that end, we have included several labs in each volume of *Top Shelf Science.*

The four years of high-school science are typically devoted to earth and space science in ninth grade, biology in tenth grade, chemistry in eleventh grade, and physics in twelfth grade. Students between grades 9 and 12 are expected to learn about modeling, evidence, organization, and measurement, and to achieve an understanding of the history of science. They should also accumulate information about scientific inquiry, especially through laboratory activity.

Our series, *Top Shelf Science,* addresses not only the national standards, but also the underlying concepts that must be understood before the national standards issues can be fully explored. National standards are addressed in specific tests for college-bound students, such as the SAT II, the ACT, and the CLEP. We hope that you will find the readings and activities useful as general information as well as in preparation for higher-level coursework and testing. For additional books in the *Top Shelf Science* series, visit our web site at walch.com.

Safety and Ethical Issues

The *Top Shelf Science* series contains several laboratory experiments. Special care must be taken to ensure student safety when performing these experiments. Experiments involving living organisms should be done with careful respect for the health of the living specimen in mind. Here are some guidelines for general safety issues in a laboratory setting:

- Wear proper safety equipment at all times. This includes an apron, a smock, or a lab coat; safety goggles; and gloves. Do not wear open-toed shoes, such as sandals, during lab experiments.
- Do not eat or drink anything in the lab.
- Be sure to turn off heat sources when not in use.
- Perform any chemical experiments involving gas emissions within a chemical fume hood, or in a well-ventilated room.
- Before disposing of chemical ingredients, be certain that they are neutralized; then dispose of them in proper containers.
- Establish a location for the disposal of sharp objects, such as broken glass or nails.
- Use extreme caution when heating solutions.
- Animals, plants, and other life forms are deserving of respect. Treat living specimens with care and, when possible, release them or replant them outdoors.
- Use care when using electrical appliances of any sort. Know how to recognize a short circuit or a blown fuse.
- Keep fire extinguishers on hand and properly charged, and know how to use them. Be sure that you have an ABC-rated extinguisher, as well as a Halon™ extinguisher for electrical fires.
- Follow all local, state, and federal safety procedures.
- Have evacuation plans clearly posted, planned, and actually tested.
- Label all containers and use original containers. Dispose of chemicals that are outdated.
- Be especially aware of the need to dispose of hazardous materials safely. Some chemistry experiments create by-products that are harmful to the environment.
- Take appropriate precautions when working with electricity. Make sure hands are dry and clean, and never touch live wires, even if connected only to a battery. Never test a battery by mouth.
- When using lasers, never look directly into the beam, and make sure students are conversant with the dangers of laser light.

Safety precautions unique to a given laboratory will be provided within the lab write-up itself. These safety precautions are provided as a guide only. They may be incomplete. Use common sense when working with any chemicals, electricity, or living organisms.

Parent/Teacher/Student Guide

Dear Parents, Teachers, and Students,

Thank you for choosing the *Top Shelf Science* series to help you better understand some of the difficult ideas in high-school science. We are confident that our books will help students who have a greater knowledge of the subject matter being studied; they can also be used to provide a lab-based program for students learning at home.

Each volume of the *Top Shelf Science* series is designed for a particular course of study. Within each volume, concepts build sequentially, and it is recommended that students begin with the first section and move forward.

Each book has sections that are thematically designed. The laboratory exercises associated with each section are specific to a deeper understanding of the overlying concept. In Appendix II, you will find a list of materials that are necessary to conduct the lab exercises, as well as a list of science equipment dealers who may help you acquire things you need in the course of the lab exercise; we have tried to keep the materials list small, as well as provide lab lessons in which materials are readily accessible. Therefore, we have also provided alternatives, where possible, to the lab glassware and other large pieces of equipment that may not be located in your kitchen cabinet or small classroom.

In Appendix I, you will find answers to the questions in each unit, as well as a suggested grading rubric for essays and lab reports. Share these rubrics with students so that they can correct areas that need to be corrected before the next assignment. In keeping with the national science standards, we have also included a time line of the history of each discipline. Each volume also contains an index and a glossary.

Whether you are using our product as the basis for a home school experience, a new and fresh way of supporting textbook material, or as preparation for a college placement test, we are confident that *Top Shelf Science* can meet your needs.

Thank you!

The authors and editors of *Top Shelf Science*

Matter and Energy

Matter and energy are two closely related concepts in chemistry that are considered in almost every chemical reaction. Everything in the universe is made of either matter or energy, and the connection between the two is one of the most compelling relationships in all of science. Scientists want to be sure that when they are making a long and complex chemical process they understand the matter and energy relationship. This enables them to predict the final products before they start.

Matter is anything that occupies space and has mass. Matter is a general term to describe the materials that make up the everyday world around us. As scientists discovered more and more kinds of substances, it became necessary to separate them into groups like solid, liquid, and gas, but all such materials are still matter.

Mass is a measure of the amount of matter in an object, usually measured in grams or kilograms. Mass is not dependent on gravity and does not change from location to location as weight does. The mass of an object can be determined in one of two ways: either it is compared to another object that has a known mass, or the weight of the object is divided by the acceleration due to gravity in the location where the object is being measured.

Everything in the universe is made of either matter or energy.

Energy is the ability to do work, usually measured in joules. Energy is found in many forms and classifications, such as kinetic, potential, nuclear, electrical, chemical, thermal, mechanical, and others.

Scientists have long thought that matter is also a form of energy, and the law of conservation of energy states that the total amount of energy in the universe is constant but can be changed from one form to another. During a chemical reaction, the amount of matter that is present at the beginning of the reaction should be the same as the amount found at the end of the reaction, and the amount of energy should also remain constant.

Exploration Activities

1. The energy stored in food is commonly measured in calories. If the SI system were used, what unit would be used to describe the amount of energy in food?

2. A box containing a piece of wood and enough air to burn the wood is measured and found to have a mass of 1.5 kilograms. If the wood is burned and none of the smoke, ashes, or hot gases are allowed to leave the box, then how will the mass of the box compare to the mass before burning? Will it be more, less, or the same?

3. The process of making light in your house from the electricity generated by a hydro-electric dam might be summarized as below:

   Mechanical energy → electrical energy → light and heat energy

   Draw your own flowchart that describes how your body turns food into the energy to walk.

BACKGROUND

Mass and Weight

In chemistry it is important to distinguish between mass and weight. Mass is a measure of the amount of matter in an object, while **weight** is the gravitational force that pulls on that amount of matter. Since weight is based on a force, its scientific unit (**newtons**) must also take into account the amount of the force. It is necessary to make measurements in science in terms of mass, because mass has an unchanging standard of measurement for a sample of matter, while weight may fluctuate from location to location.

Mass has an unchanging standard of measurement for a sample of matter, while weight may fluctuate from location to location.

One of the reasons we use mass in chemistry is that it allows us to keep track of the amount of matter in any situation in which experimentation or measurement is being used. The reactants in a chemical reaction combine in very specific quantities, and those quantities are directly related to the amount of each reactant that is present. Although the weight might change from location to location, the ratios that form in these reactions are directly related to their mass, regardless of their location. This would apply to a reaction or a measurement made on the equator, the Moon, or in orbit around Earth, where the effects of gravity are nullified by free fall.

For example, an astronaut with a mass of 100 kilograms will have a weight of approximately 980 newtons on Earth, while on the Moon her weight would only be about 160 newtons. Since newtons are a measure of force, this shows that the amount of force a person would exert on a pair of scales on the moon is almost $\frac{1}{6}$ of what it would be on Earth, even though the amount of matter in the person has not changed. Even while in free fall, they would have a weight of 0 newtons and their mass would still be 100 kilograms. If you find this hard to believe, imagine that you are in the cargo bay of the space shuttle and one of the astronauts pushes a full-sized car across the bay and it hits you moving at 20 mph. Despite the fact that it is weightless, it would hit you just as hard as it would on Earth because it has the same amount of material and the same amount of **inertia.**

Exploration Activities

1. Name one product that people commonly purchase by mass and not by weight.

2. Which would contain more matter, 500 kilograms on the Moon or 500 kilograms on Jupiter?

3. Which would weigh more, 500 kilograms on the Moon or 500 kilograms on Jupiter?

BACKGROUND

States of Matter

Chemistry has been called the science of identification. A chemist often spends a large amount of time classifying substances and determining the ways those substances interact with one another. To identify a substance, one of the first things a chemist could do is determine the material's state. The **state of matter** is the physical condition of a substance, such as solid, liquid, or gas.

The state of matter is the physical condition of a substance, such as solid, liquid, or gas.

States of Matter

Solid is the state of matter that is rigid, in which the particles are close together and the material has a fixed shape and volume. The solid form of water, ice, is a simple example.

Liquid is the fluid state of matter in which particles are close together; a liquid has a definite volume and an indefinite shape. At room temperature, water, alcohol, and cooking oil are all liquids.

Gas is the fluid state of matter in which the molecules are widely spread apart; a gas has neither a definite shape nor volume. The air we breathe is filled with such gases as oxygen, nitrogen, and argon.

Plasma is the fourth state of matter, often characterized as an ionized gas. Some plasmas are "cool" plasmas, such as the gas inside of a fluorescent light bulb as electricity travels through it, or "hot" plasmas, such as stars. When a simple substance gets too hot, its individual components may actually be torn apart by collisions with other particles, leaving behind only charged particles.

For any single substance, such as water, there is an increase of kinetic energy, and therefore temperature, as the material changes from ice to water to steam. Although the steam can still be heated and its temperature can still increase, even if it gets hot enough to become a plasma, there is no special name for water in this phase. The individual molecules that make up water would most likely be separated into their component parts at such high temperatures, and consequently the material would be an ionized gas, not just another state of pure water.

More on Solids

At the atomic level, the atoms and molecules that make up solids stay in one place, and although they vibrate where they are, they essentially do not move from their location. This internal structure of many solids is made up of crystals. A **crystal** is a structure with plane faces that have a repeating geometric pattern, and solids that have this kind of internal structure are called crystalline solids. The repeating pattern of atoms in these substances is called a **crystal lattice.** The crystalline nature of some substances is easily seen, as it is in salt or gemstones such as diamonds. In other materials, like steel, it can only be seen under high magnification, but the patterns are there.

There is another group of solids that are called amorphous solids. An **amorphous solid** is one in which the atoms or molecules that make up the solid are in a disorganized arrangement. Some examples of amorphous solids are rubber, glass, and paper.

Sublimation is the change of a solid to a vapor without passing through the liquid phase.

When energy, usually in the form of heat, is added to many solids, the particles increase their vibration as their kinetic energy increases, until they reach a point at which they do more than vibrate. In fact, when the particles in a solid get enough energy, they are able to leave their usually stable locations and flow around one another. The point at which this takes place is called the **melting point,** and it is the temperature at which a substance changes from a solid to a liquid. When a substance is already a liquid and it loses enough energy to change back to a solid, this point is called the freezing point. The freezing point and melting point are the same temperature for a given substance, but what determines whether the substance will melt or freeze is whether or not it is gaining or losing energy.

One other footnote about some solids is that they can skip the liquid phase when they melt and become a vapor. **Sublimation** is the change of a solid to a vapor without passing through the liquid phase. Dry ice, which is the solid state of carbon dioxide, sublimates at –78.5°C, which makes it great as a method for keeping food cold. It can keep foods frozen since they only have to be below 0°C, and when the dry ice finally warms up enough to melt, it doesn't. Instead, it turns to a gas, making for a material that doesn't get the food wet.

More on Liquids

When particles in the liquid phase gain enough kinetic energy, they turn to a vapor. This can happen either at the boiling point, or through evaporation. The major difference between the two is that the **boiling point** is a specific temperature, and boiling can happen at any location throughout the liquid that is being heated. **Evaporation** occurs when particles in the liquid phase escape the liquid and enter the gas phase, usually at a temperature other than the boiling point, and only when the surface of the liquid touches the atmosphere around it.

Distillation is the process of vaporizing and condensing a substance or mixture to purify or separate its various components.

Each substance that can be boiled has a very specific boiling point. When a collection of liquids are mixed together, they can still be separated by distillation. **Distillation** is the process of vaporizing and condensing a substance or mixture to purify or separate its various components. As a mixture is heated and its temperature increases, eventually the temperature will reach the boiling point of the substance that has the lowest boiling point. When that happens the temperature of the mixture will not increase until that substance has boiled off. After the released gases are captured and condensed while the temperature is constant, the temperature will continue to rise until it reaches the next boiling point, and the process can be repeated until all substances are separated out.

Another property of many liquids is surface tension. **Surface tension** is the tendency of molecules at the surface of a liquid to be pulled together. Typically this is caused by an imbalance in inter-molecular forces. If you think of the surface of water as being made of a sheet of water molecules, that sheet has a greater attraction for the water molecules under it than it does for the particles of air that are over it. When the sheet has a greater attraction for the surface it is contacting than for the water under it, we see **capillary action**—the tendency of a liquid surface to follow a surface it is in contact with because of **adhesion** and surface tension.

When we see the liquid simply by itself, the attraction between the molecules in the liquid results in viscosity. **Viscosity** is the resistance of a fluid to flow. Hydrogen-bonded liquids usually have high viscosities—that is, they are "thick" and don't flow very well. For example, maple syrup at room temperature has a higher viscosity than water.

Exploration Activities

1. The air around us contains the gaseous form of nitrogen. If the nitrogen were cooled enough to change state, what state would it be in, solid, liquid, or gas?

2. Rubbing alcohol is a liquid at room temperature. If it were cooled enough to change state, what state would it be in, solid, liquid, or gas?

3. Can you name a substance other than water that you have seen in solid, liquid, and vapor form? If so, what was it?

4. Define "crystal."

5. You have a solution of water and rubbing alcohol. How could you separate them?

Student Lab: Change of Phase

As many pure substances change from one phase to another, their temperature stops changing while energy is used to rearrange the internal structure of the material. It also allows us to quite accurately determine the melting and boiling points of a material. In this lab you will find the plateaus on a graph of temperature versus time and try to accurately determine the boiling point and melting points for water.

Materials

- Thermometer
- Ice
- 600-mL beaker
- Hot plate

Safety Considerations

Students should wear goggles and lab aprons. Thermometers should be very carefully used to stir the water-ice mixture. The hot plate will eventually get very hot, as will the beaker and its materials. Students should not handle the setup until it has had ample time to cool at the end of the lab.

Procedure

1. Fill the 600-mL beaker about two-thirds with ice.
2. Record the temperature of the ice by placing the thermometer in the center and leaving it for approximately 3 minutes.
3. Record the temperature every 30 seconds until the end of the lab.
4. Place the beaker on the hot plate and turn on the heat.
5. Record the temperature until the water boils, and when it seems to be boiling continue recording for 3 minutes to make sure the temperature is not increasing.
6. When the temperature has been the same for at least three readings after boiling is reached, stop recording temperatures.

Comprehension Questions

1. Create a graph of temperature versus time.

2. Are there any locations on the graph where the line is essentially horizontal? What does this indicate?

3. What is the melting point of water according to your graph?

4. What is the boiling point of water according to your graph?

5. The accepted melting point of water is 0°C and the accepted boiling point is 100°C. If your numbers were different, what might account for the inconsistency?

Chemical and Physical Properties

> **Density, color, odor, melting point, boiling point, taste, electrical conductivity, and hardness are all examples of physical properties.**

In the continuing process of learning to identify the various types of substances, it becomes important to be able to separate pure substances from mixtures. One of the primary methods for distinguishing among the various types of matter is to use the properties that can be observed. One of the first things a scientist would use when identifying a substance is a list of physical properties. The **physical properties** are the characteristics of a substance that can be observed without the production of new chemical materials. Density, color, odor, melting point, boiling point, taste, electrical conductivity, and hardness are all examples of properties that can be observed in a substance without changing the chemical makeup of the material.

After the physical properties of a substance are recorded, a scientist will often go on to try to identify the chemical properties. The **chemical properties** are those characteristics of a substance that are observed as the substance undergoes a chemical change. In the process of observing the chemical properties of a substance, the scientist will often use small samples of the material and will chemically alter them in the testing process. Finding the answers to a list of questions might be a way to test the chemical properties of a substance: Will it burn? Will it react with acids and/or bases? How does it interact with other known substances? Does it react with metals (**corrosion**)? As each question is answered, the experimenter is able to narrow down the list of possible substances by comparing the results to substances of known composition.

A chemist may be required to identify a material that is a clear, colorless liquid. It might be easy to assume that the substance is water, but rubbing alcohol, methanol, hydrochloric acid, and ethanol are also liquids that fit the basic description. If the material were flammable, then water could be ruled out. Trying to freeze or boil the liquid would allow it to be identified by checking a list of boiling and melting points for various substances and making a comparison. Other properties such as density or an ability to interact with acids, bases, or metals would narrow down the possible substances even further. Even if there are two or three substances that dissolve metal and have similar boiling points, it is unlikely that they all have the same density and viscosity as well.

Exploration Activities

1. Identify each of the following as chemical or physical properties:

   a. Hardness

   b. Flammability

   c. Density

   d. Reactivity with acids

   e. Reactivity with metals

2. Name all of the physical and chemical properties that you can think of to describe the following:

   a. Water

   b. Wood

   c. Iron

   d. Chocolate

3. Is there a good reason to start with physical properties instead of chemical properties when identifying a substance? Why or why not?

Chemical and Physical Changes

> **Ripping a sheet of paper in half, crushing a soda can, chopping up lettuce for a salad, and boiling water are all examples of physical changes.**

In the process of determining the physical and chemical properties of a substance, the experimenter may chose to put the substance through a variety of changes. Some of these changes will be physical in nature, while others will be chemical. In testing the physical properties of a substance, the material may be put through a **physical change,** which occurs when the physical properties of the substance, but not the chemical properties, are changed. If a pencil were broken in half, no new chemicals would be produced. That is to say, the chemicals that were present before the break would still be present in the same proportions. Ripping a sheet of paper in half, crushing a soda can, chopping up lettuce for a salad, and boiling water are all examples of physical changes. Often it has been said that, in a perfect lab, with perfect equipment and technique, a physical change can be reversed without the use of chemicals.

In testing the chemical properties of a substance, however, the material being tested will interact with other substances in a manner that produces different substances. A **chemical change** is the result of a chemical reaction, during which there is a change of the properties and composition of a substance. Products are made that have different properties than those of the original substances. When a piece of wood is burned, the result is ashes, soot, light, heat, gases, and even water vapor. None of these things looks or behaves like wood, but each is a by-product of the burning. Likewise the rusting of steel, the corrosion of metal by acid, the toasting of a piece of bread, the dissolving of hard water deposits from a sink, or the burning of a candle are all situations in which a chemical change is taking place.

Scientists often use the most basic of lab equipment to test the properties of matter. A scale, graduated cylinder, and water are all that is needed to calculate the density of a solid, assuming it doesn't dissolve in water. Many of the physical and chemical changes that occur in a material can be observed with the five senses without the use of any special equipment. It is no stretch of the imagination to see that if a piece of chalk is crushed it is undergoing a physical change. Likewise, when baking soda and vinegar are mixed together it is not hard to see that the bubbles coming out of the solution constitute a material that was not present when the baking soda and vinegar were still individual materials.

Exploration Activities

1. Name five examples of a physical change.

2. Name five examples of a chemical change.

3. Describe a situation in which you could see a physical change and a chemical change in the same material.

Categories of Material

There are millions of chemicals and substances that are used in industry, found in nature, and sought after by scientists. It is impossible for any one person to be able to identify every possible material simply by looking at it or from memory. Over the many years that scientists have collected information about the environment we live in, they have amassed a vast amount of data that can be used to identify the materials we find anywhere on Earth. They have spent a lot of time classifying substances and creating methods of identifying even the most complex material. One of the ways they have done this is to create a system in which substances can easily be categorized and then placed into subcategories.

A substance is a pure sample of matter, either an element or a compound.

The first category is the **element**, which is a substance made out of one kind of atom that cannot be separated into simpler substances by ordinary chemical means. All of the atoms in a sample of copper are copper atoms. They have the same atomic number and all exhibit the properties of copper. A list of the elements can be found on a **periodic table**, and although there are a large number of naturally occurring elements, there are also some that are synthetic—that is, they do not occur naturally and are made by human beings.

The next category is the **compound**, which is a substance made of atoms of two or more elements combined in a definite proportion. For example, water is a compound of hydrogen and oxygen, and sugar is a compound of carbon, hydrogen, and oxygen. Compounds are usually classified as **organic** or **inorganic.** Organic means that the compound contains the element carbon and, usually, hydrogen.

A **substance** is a pure sample of matter, either an element or a compound. This is a more general phrase, and can only be applied to materials that have a single, distinct set of properties. For example, water is a pure substance, while soda is not. The soda might contain water, sugar, caffeine, and phosphoric acid, all of which are pure substances that have been put together in a complex mixture. Many times the list of ingredients on a product such as deodorant or a bag of candy is actually a list of the pure substances to be found in the product.

Finally, a **mixture** is a combination of two or more substances that do not chemically combine and can be separated by physical means. In terms of elements and compounds there are three ways a mixture could be combined:

1. A mixture could be made of two elements, like iron pellets and copper pellets mixed in a bowl.
2. A mixture could be of two kinds of compounds combined, like sugar and cinnamon in a single shaker.
3. A mixture could be an element and compound combined, like aluminum pellets and water.

There are two major kinds of mixtures. Heterogeneous mixtures have components that are not evenly mixed throughout. For example, when sampled from different locations throughout, the heterogeneous mixture of a salad might produce lettuce in one location and tomato in the next. Likewise, cookie dough ice cream sometimes has areas that are high in cookie dough chunks, while other parts are only vanilla ice cream.

Homogeneous mixtures, on the other hand, have their components evenly distributed throughout. A glass of unsaturated salt water is a good example. The amount of salt in a sample at the top of the glass is the same as the amount of salt that would be found in an equally-sized sample from the bottom of the glass.

Exploration Activities

Identify each of the following as an element, compound, substance, or mixture. Some examples may have more than one answer.

1. toast
2. peanut butter
3. ice
4. cheese
5. brass
6. sulfur

Units of Measurement

SI (Système International d'Unités)

The **SI (Système International d'Unités),** or International System of Units, is a system of basic units for measurement that is the most modern form of what started as the metric system in the 1790s. The SI system has a set of base units that make up the foundation for the most common forms of measurement that scientists use. These units are seen in the chart below.

Combinations of base units are called derived units.

SI Base Units

Quantity	Base Unit
Chemical amount	1 mole (1 mol)
Electric current	1 ampere (1 A)
Length	1 meter (1 m)
Luminous intensity	1 candela (1 cd)
Mass	1 kilogram (1 kg)
Temperature	1 Kelvin (1 K)
Time	1 second (1 s)

Combinations of these base units are called **derived units.** You can tell a unit is derived when its label contains more than one of the base units. For example, velocity is a measure of the rate of motion of an object. This requires both time and length, so it is measured in meters per second, or m/s.

Significant Figures

Significant figures are the digits in a measurement that are certain, including one digit in the position of least value that is uncertain. For example, when measuring a line with a ruler, it is easy to see that it is between 1.1 and 1.2 cm, but where exactly does the line end? Some people might say exactly halfway between, making the answer 1.15 cm, but some people might see it as a little more or less, perhaps 1.14 or 1.16 cm. The estimation of the last place makes that digit uncertain. Typically, this means that when data is given as a number such as 4.39, it could actually be 4.38 or 4.40, unless the amount of uncertainty is indicated to be otherwise.

The Zero Rules

1. Measured zeros are significant.
2. Zeros as placeholders are not significant unless otherwise indicated. 0.00023 only has two significant digits because the zeros are placeholders.
3. Zeros surrounded by nonzero numbers are significant. 405 has three significant figures because the zero is between two nonzero digits.
4. Placeholding zeros to the left of a decimal point are significant if there are also nonzero digits to the left of the decimal point. 1,400 has two significant figures, unless otherwise indicated, because no decimal is shown, while 1,400. (note the period that is not followed by zeros) has four significant digits.

EXPLORE

Exploration Activities

1. Name a combined unit you have seen in everyday life.

2. Why is it important to have an international system of measurement?

3. How many significant figures are there in each of the following?

    a. 4.1

    b. 4.10

    c. 45.01

    d. 1,300

    e. 14,300.

    f. 0.00203

Density and Specific Gravity

One of the most useful methods of identifying matter is to determine its density. **Density** is the measure of the amount of mass in a given volume. Although the material that makes up an empty soda can is not destroyed when it is crushed, its density increases. Density equals mass divided by volume or is given by the formula $D = m/V$ where

Density equals mass divided by volume.

D = density measured in g/cm^3 or g/mL
m = mass in grams
V = volume in cm^3 or milliliters

A further useful comparison is specific gravity. **Specific gravity** is the ratio of the mass of a given volume of any substance to a mass of an equal volume of water usually at 4°C. Gold has a specific gravity of around 19.3. This means that if an equal volume of gold and water were compared, then the mass of the gold would be 19.3 times the mass of the water. Sometimes specific gravity is calculated by using the known density of a material and the density of water. It is a ratio where the density of the material is divided by the density of water. For example, the density of mercury is 13.55 g/cm^3 at 20°C and the density of water is 1.00 g/cm^3. When 13.55 g/cm^3 is divided by 1.00 g/cm^3, the value of 13.55 does not change, but the labels are canceled leaving a number with no units. This is not unusual because ratios by definition have no units.

Exploration Activities

1. What is the density of a 45-g block of wood that occupies 38.2 cm^3?

2. What is the specific gravity of copper if its density is 8.96 g/cm^3?

3. If a piece of wood has a specific gravity of 0.45, will it float on pure water?

Heat and Temperature

Heat can be transferred from a warm body to another body at lower temperature by the processes of radiation, conduction, and convection.

Heat is a form of energy that can be transferred between two bodies of differing temperature. Heat can be transferred from a warm body to another body at lower temperature by the processes of **radiation, conduction,** and **convection.** Radiation is the transmission of heat through waves, rays, or particles. Conduction is the transfer of heat by direct contact. Convection occurs when heat is passed to a fluid around a hot body and then the heat is carried to another body by the fluid.

Temperature is the measure of the average kinetic energy in the particles of a sample of matter. This is often confused with heat, but heat is a form of energy, while temperature is a way of measuring the amount of heat energy in a substance. As the kinetic energy of a substance increases, the speed of its particles increases, and the reading of a thermometer will also increase as some of the kinetic energy is passed to the measuring device.

An **exothermic** reaction is a chemical reaction or process that releases heat. Any process during which a material or reaction loses heat is called exothermic. Burning a piece of wood, burning a sheet of paper, or mixing sulfuric acid with water are all interactions that will produce heat and are thus exothermic.

Endothermic reactions are chemical reactions or processes that absorb heat. The process of melting ice is endothermic as heat from the surrounding environment is absorbed by the ice to change its phase.

Exploration Activities

1. Is the changing of water into steam endothermic or exothermic?

2. Name as many places as you can in a house where you would find examples of conduction.

3. Name as many places as you can in a house where you would find examples of convection.

Student Lab: Physical Properties of Compounds

Each person has a set of skills and senses that allow him or her to observe the environment. We use our senses to identify milk that has soured, pick out matching socks, and determine the location of a friend who is yelling to us from the other side of a crowded room. We like to be able to identify everything around us, and we do so all the time without thinking. Just as you can tell that a particular school book is yours by the way you have doodled on the cover, we can also determine the makeup of substances we are unfamiliar with by categorizing all the things we can observe about it.

Materials

- Compounds to be tested: salt, vegetable oil, and sugar
- Beaker
- Electrical conductivity probe
- Glass stirring rod
- Graduated cylinder
- Hydrogen peroxide
- Scale
- Water

Safety Considerations

Goggles and lab aprons should be worn.

Students should not taste any of the materials in the lab, despite the fact that they are common household items.

Procedure

Repeat all steps of the procedure for each compound one at a time.

1. Describe the physical appearance of the salt, including color, odor, state at room temperature, and any other properties you can think of that can be tested with the five senses (except taste!).

2. Fill a clean, small beaker (100 milliliters will do) about half full with hydrogen peroxide. Sprinkle a little salt into the beaker and stir with a glass stirring rod. Is the material soluble in hydrogen peroxide?
3. Fill a clean, small beaker about half full with water, and sprinkle a little salt into the water and stir with the glass stirring rod. Is the salt soluble in water?
4. Record the mass of a dry, clean graduated cylinder. Now add approximately 10 milliliters of salt to the cylinder and record the mass of salt and cylinder. Now, calculate the density of the salt.
5. Pour some salt into a ceramic bowl, pan, or evaporating dish. Insert the conductivity probe into the salt. Does it conduct electricity?
6. Perform steps 1–5 with the cooking oil, and then with sugar.

Comprehension Questions

1. In what major ways were cooking oil and salt different from each other?
2. In what ways were salt and sugar similar to each other?
3. Did any of the tests in the procedure reveal a difference between salt and sugar?
4. What is one way you could tell the difference between salt and sugar that is not generally considered safe lab practice?
5. Hydrogen peroxide is a nonpolar liquid and water is a polar liquid. A polar liquid will dissolve a polar substance, and a nonpolar liquid will dissolve a nonpolar substance. Based on this fact, which of the test samples, salt, vegetable oil, and sugar, were polar and which were nonpolar?

Student Lab: Chemical Properties of Substances

The nature of a single material is often easy to determine simply by looking at that material, especially when it is in a familiar context. When a clear, colorless, odorless liquid comes out of a faucet in your kitchen, it is usually safe to think that it is water. However, a clear, colorless, odorless liquid found in a beaker in a science lab could be any number of things, including some that are hazardous to human health. Knowing the exact nature of such a material is important for safety, and when we understand the exact nature of a substance we may be able to find more uses for it in everyday life. One of the best ways to do this is to expose the material to chemicals, electricity, or other forms of stimulation and observe how it reacts. In this lab you will observe, as carefully as possible, the behavior of a material under several different conditions.

Materials

- Part 1: hydrogen peroxide, raw potato, and a 250–600 mL beaker
- Part 2: iron powder, sulfur powder, a magnet, and a large beaker or evaporating dish

Safety Considerations

Goggles and lab aprons should be worn.

The heating component of this lab should be done with proper ventilation or in a fume hood.

Procedure

Part 1

1. Fill the beaker with enough hydrogen peroxide to cover your slice of potato.
2. Add the slice of raw potato.
3. Watch the surface of the potato until it is clear whether a reaction is taking place, and record your observations.

Part 2

1. Combine about 15 grams of iron powder with about 30 grams of sulfur in an evaporating dish.
2. Wrap the magnet in a piece of notebook paper, only one layer thick, and remove the iron powder by rubbing the paper-wrapped magnet in the mixture. Remove the paper and let the iron fall back into the sulfur.
3. Mix the sulfur and iron together and heat using a Bunsen burner with a low flame. (*Caution:* This should be done in a fume hood or in an area with excellent ventilation.)
4. After the mixture has combined into a single substance and cooled off, try to separate the iron out again.

Comprehension Questions

1. What did you observe that indicated the potato was undergoing a chemical change?

2. What did you observe that indicated the iron and sulfur were undergoing a physical change?

3. Why were you able to separate the iron from the mixture before it was heated, but not after?

Student Lab: Determining Density

Density is a fairly simple way of telling one material from another. Hydrogen peroxide, rubbing alcohol, and water all look very similar, but they have very different properties. Although you could identify these substances by their odors, that isn't a particularly safe way of identifying an unknown material. One safer way of identifying any substance is by its density, and in this lab you will learn how to calculate the volume of regular and irregular objects.

Materials

- One small cube of any material that doesn't float in water
- Graduated cylinder the above cube could easily fit into
- Five nickels
- Scales or digital balance
- Ruler

Procedure

1. Measure the length, width, and height of the cube.
2. Measure the mass of the cube.
3. Fill the graduated cylinder about half full with water and record the level. Carefully lower the cube into the water so it doesn't splash and let it sink to the bottom. Try not to get your fingers in the water.
4. Record the new level of the water.
5. Repeat steps 1–4 with the five nickels.

Calculations

1. Use the formula for volume of a cube ($l \times w \times h$) to calculate the volume of the cube.
2. Calculate the volume of the cube by finding the difference between the water level with and without the cube.
3. Calculate the density of the cube, using the formula $D = m/V$. Use the volume calculated in step 1.

4. Calculate the density of the cube using the volume you learned from the calculation in step 2.

5. Calculate the volume of the nickels by using the formula $V = \pi r^2 h$ (in which r equals the radius and h equals the height).

6. Calculate the volume of the nickels by finding the difference between the water level with and without the nickels.

7. Calculate the density of the nickels by using the volume you learned from the calculation in step 5.

8. Calculate the density of the nickels using the volume you learned from the calculation in step 6.

Comprehension Questions

1. Could the volume of an irregularly shaped object, such as a rock, be easily calculated by measuring the object with a ruler and using a volume formula? Why or why not?

2. How closely did your two densities compare? Which method do you think was the most accurate?

3. How would you calculate the density of a cylinder of brass?

BACKGROUND

Atoms and Molecules

As we identify and study all of the types of matter, it becomes more important to determine the individual components that are found at the smallest levels of that matter. Scientists had long believed that there were small particles that were the building blocks of matter, and it is now possible to see images of these particles, which are called atoms.

The Greek philosopher Democritus put forward one of the first coherent theories about the existence of **atoms** nearly 2,400 years ago. In part it states:

1. Atoms are indivisible, solid, and indestructible.
2. Matter is made of mostly empty space, and atoms move through the space.
3. Atoms of different kinds have different shapes and sizes.
4. The differing properties of atoms are related to their different shapes and sizes.
5. When matter changes it is because of a rearrangement of atoms, not because the atoms themselves change.

It is amazing how close Democritus was in his ideas, but it is helpful to remember that he was working with ideas and not the scientific method. Critics in his own time were able to derail his train of thought simply by asking questions like "What holds the atoms together?" and "How can there be 'empty' space?" Proper scientific research requires that such thoughts be followed up with experimentation and modified accordingly when the data supports or denies the ideas put forward.

An atom of gold is different from an atom of any other element.

Atoms are the smallest particles of an element that contain all the properties of that element. An atom of gold is different from an atom of any other element in that it will have a different number of protons and electrons, and it may have a different number of neutrons.

A **molecule** is the smallest particle of a compound that maintains all the properties of that compound. Molecules are typically made of two or more atoms that are bonded together and are electrically neutral.

Our understanding of how atoms combine to form molecules is largely thanks to the work of John Dalton, an English chemist who worked in the late 1700s and early 1800s. He created an atomic theory that has been changed only slightly in modern times.

1. Elements are made of atoms, which are indivisible. (We now know that atoms are made of **protons, neutrons,** and **electrons,** and even smaller particles, and can be divided.)

2. All atoms of an element are the same. This would be true except for the existence of **isotopes,** which are atoms of an element that have different atomic masses because they have differing numbers of neutrons. However, the *average* mass of the isotopes in a naturally occurring sample of matter is roughly constant and different from element to element.

3. Atoms of different elements are different because they have different masses. This is true today, although we are more likely to distinguish them by saying they have different numbers of protons.

4. Compounds are formed by atoms joining together in a manner such that the number of each kind of atom can be defined by a distinct whole-number ratio. For example, H_2O is water, while H_2O_2 is hydrogen peroxide. While the first has a ratio of 2:1, the second has a ratio of 2:2, which allows us to distinguish between compounds that have the same kinds of atoms but in different ratios.

Exploration Activities

1. Why was Dalton wrong in thinking that atoms could not be divided?

2. How can we tell that CO (carbon monoxide) is different from CO_2 (carbon dioxide) if they both contain carbon and oxygen?

3. What is an isotope?

Chemical Formulas

When using a substance like carbon monoxide, it is very time-consuming to write out the name over and over when describing a chemical reaction, so scientists have devised a shorthand method of writing the names of the various chemical substances. A **chemical formula** is a shorthand method using numbers and chemical symbols to describe the makeup of one molecule of a substance. Chemical symbols are the abbreviations found for each element on the periodic table. For example, oxygen is represented by O, calcium by Ca, and iron by Fe. Every symbol consists of at least a single capital letter, and some are followed by a second lowercase letter. Some of the symbols are based on what we think of as the "English" name—such as lithium, symbolized by Li—while others are based on other languages like Latin and German. For example, the element mercury is represented by Hg, which is from the Latin word *hydragyrum,* roughly translated as "liquid silver."

Chemical symbols are the abbreviations found for each element on the periodic table.

These elemental symbols are combined together, with numbers that show the ratio of each element in a compound, as a shorter method of writing out a molecule's name. Some examples:

Chemical Name	Formula	Number of Atoms Represented
Carbon monoxide	CO	One carbon, one oxygen
Sodium chloride	NaCl	One sodium, one chlorine
Water	H_2O	Two hydrogen, one oxygen
Ammonia	NH_3	One nitrogen, three hydrogen
Hydrogen gas	H_2	Two hydrogen

Exploration Activities

1. Find a copy of the periodic table of elements on the web, and identify the elements in each of the following compounds:

   a. H_2SO_4

   b. $BaCl_3$

   c. CH_4

2. Write the formula for each of the following compounds given the elements and the ratios in which they occur:

   a. Lithium, oxygen—2:1

   b. Carbon, chlorine—1:4

   c. Silver, phosphorus, oxygen—3:1:4

Ions

Certain chemical substances are held together by the strong electrostatic attraction between the ions that are in the substance. An **ion** is an atom or a group of atoms that has acquired a charge due to a loss or gain of electrons. An atom with more electrons than protons will have a negative charge, just as an atom with more protons than electrons will have a positive charge. When the attraction between these positive and negative ions is strong enough, it will allow the ions to be bonded together to form a single substance. For example, lithium is an element that tends to lose an electron easily and, as an ion usually has a charge of positive one, is symbolized by Li^{1+}. Likewise, oxygen is an element that easily gains two electrons and has a charge of negative two, symbolized by O^{2-}.

An atom with more electrons than protons will have a negative charge; an atom with more protons than electrons will have a positive charge.

A **polyatomic ion** is a group of two or more elements bonded together and having a collective charge. Sulfur and oxygen often bond together in such an arrangement called sulfate. The sulfate ion has a charge of negative two and is symbolized as SO_4^{2-}.

Different combinations of positive and negative ions produce different substances, and when the bonding process is finished, ionic compounds will have combined in such a manner that they will be electrically neutral. That is, they will have a collective charge of zero. Look at the reactions below:

$$Na^{1+} + Cl^{1-} \rightarrow NaCl$$

or

$$K^{1+} + K^{1+} + SO_3^{2-} \rightarrow K_2SO_3$$

Each of these reactions has an equal number of positive and negative charges so that when the positive and negative ions are combined, the sum of their charges is zero. There is an extensive list of ions and their charges in Appendix II.

Exploration Activities

1. What is the charge of the sodium ion in the formula Na_2O?

2. What is the collective charge of H_2O?

3. What would the formula be for a compound made from lithium and oxygen?

Atomic Mass

When differentiating between kinds of atoms, it becomes important to be able to tell them apart by their masses. **Atomic mass** is the mass of an atom as measured in atomic mass units. Although many standards for calculating the mass of atoms were used in the past, currently scientists use the **atomic mass unit,** which is the unit of mass equal to $^1/_{12}$ the mass of a carbon-12 atom. The carbon-12 atom is an atom that has six protons, six neutrons, and 12 electrons, making in total a mass of 1.9926×10^{-23} g. This means that $^1/_{12}$ of that mass, or one atomic mass unit, is equal to 1.6605×10^{-24} g. The atomic mass unit is often abbreviated as amu, and more recently as just the letter u. The atomic mass of each element is given on the periodic table, and is actually a weighted average of all the naturally occurring isotopes of an element. For example, there are isotopes of carbon-13 and carbon-14, but because they occur in much smaller amounts than carbon-12, they don't change the atomic mass of carbon very much. However, the mass of carbon is listed as 12.0111 u and not exactly 12 u.

Scientists use the atomic mass unit as the standard measure for calculating the mass of atoms.

Molecular Mass

Molecular mass is the mass of one molecule of a substance, given in atomic mass units. For example, to calculate the molecular mass of a compound like water, which has a formula of H_2O, one would add the atomic masses of two hydrogen atoms and one oxygen atom. 1.00794 u + 1.00794 u +15.9994 u = 18.01528 u. This procedure is a little more detailed than the one that would be used in the typical high school lab. Typically, when taking atomic masses from the periodic table, your teacher will only have you use the values out to the tenths or hundredths place.

Avogadro's Number

Avogadro's number, named after Italian scientist Amedeo Avogadro, is the number of atoms in 1 mole of a substance, equal to 6.02×10^{23}. Knowing this number allows us to calculate the number of particles of a substance needed to have a numerically equal value of atomic mass or formula mass to the gram atomic mass or gram formula mass. In other words, if the atomic mass of oxygen is

16.0 u, then Avogadro's number of oxygen atoms would have a mass of 16.0 grams. This number of particles is also referred to as the mole. The **mole** is the SI unit of amount of matter, often given as 6.02×10^{23} particles (atoms, molecules, ions) of a substance. The mole can be thought of much the same way as a dozen is. A dozen eggs, a dozen pencils, or a dozen atoms all have a numerical value of 12. Likewise, a mole of eggs, a mole of pencils, a mole of atoms, or a mole of molecules would be 6.02×10^{23} of each of those items.

If the atomic mass of one oxygen atom is 16.0 u, and the mass of 1 u in grams is 1.6605×10^{-24} g, then the mass of one oxygen atom is $(16.0\ u)(1.6605 \times 10^{-24}\ g/u) = 2.6568 \times 10^{-23}$ g. Therefore, if there is one mole of oxygen atoms, then the total mass is $(6.02 \times 10^{23})(2.6568 \times 10^{-23}\ g) = 16.0$ g.

Exploration Activities

1. How many carbon atoms are there in 1 mole of carbon atoms?

2. What is the molecular mass of glucose, $C_6H_{12}O_6$?

3. What is the mass of a phosphorus atom in grams? In atomic mass units?

Student Lab: Determining Molar Mass

The **molar mass** of a material is one way of identifying a material. It also allows us to tell the difference between hydrogen peroxide (H_2O_2) and water (H_2O). Even though these two substances are made out of hydrogen and oxygen, the exact ratio by which they combine makes them quite different.

Materials

- Disposable butane lighter
- 1-liter graduated cylinder
- Digital scale or lab balance
- 1-gallon or larger bucket
- Thermometer
- Plastic wrap
- Barometer or access to barometric pressure reading for your local area

Safety Considerations

Butane is highly flammable and mildly toxic. There should be no open flames in the lab, and if possible the flints or rollers should be removed from the lighters before using, to prevent accidental striking.

The gas should be properly disposed of outside or in a fume hood.

Goggles should be worn at all times.

Procedure

1. Measure the mass of the full butane lighter.
2. Submerge the 1-liter graduated cylinder in water and let it fill.
3. Turn the cylinder upside down so that the water stays inside.
4. Submerge the butane lighter under the cylinder so that when the gas release is pushed the gas bubbles will collect in the cylinder.

5. Fill the cylinder with about 800 milliliters of butane gas, and leave under the water with the gas still in it.

6. Read the amount of butane gas in the cylinder carefully and record it. Leave the cylinder under water with the gas still in it.

7. Cover the opening to the cylinder with a piece of plastic wrap and remove from the water. Be careful not to let the gas escape.

8. Dispose of the gas outside or in a fume hood.

9. When the gas has been properly disposed of, dry the butane lighter, possibly with a hair dryer or fan, and measure its mass again.

Data

Initial mass of lighter (g) ____________

Final mass of lighter (g) ____________

Volume of gas collected (ml) ____________

Barometric pressure (atm, kPa, torr) ____________

Water temperature (°C) ____________

Molar mass of butane (g/mol) ____________

Calculations

Show your work.

1. Calculate the volume of gas collected in liters.

2. Convert your water temperature into Kelvin.

3. Calculate the mass of the gas (initial mass – final mass).

4. Convert the volume of the gas to standard temperature and pressure (STP) by correcting for pressure and temperature.

5. Calculate the density of butane gas at STP.

6. Calculate the molar mass of butane.

7. The accepted value for the molar mass of butane is 58.0 g/mol. Calculate your percent error.

Comprehension Questions

1. The barometric pressure must be corrected for the vapor pressure of water. Do this adjustment and indicate how it changed your percent error.

2. What is one source of lab error? (Don't say human error, this is always vague and obvious unless the lab is being done by monkeys!)

Skill Builder: Calculating Formula Mass

Example: Each material has its own particular mass that depends on the masses of the atoms that make up the substance. Common table sugar, also called sucrose, has a formula of $C_{12}H_{22}O_{11}$. To calculate its mass you would have to add the mass of 12 carbon atoms, 22 hydrogen atoms, and 11 oxygen atoms.

(12 carbons × 12.01 u) + (22 hydrogen × 1.01 u) + (11 oxygen × 16.00 u) = 342.34 u

Instructions

Find the molecular or formula mass for each of the substances listed below.

1. NaCl
2. H_2O
3. KCl
4. MgO_2
5. $BaCO_3$
6. Fe_2O_3
7. P_4O_{10}
8. CH_4
9. C_5H_{12}
10. C_6H_5OH
11. NH_4Cl
12. $AgNO_3$
13. $CaCO_3$
14. BF_3
15. $Ca(OH)_2$
16. HCl
17. $PbSO_4$
18. $MgCO_3$
19. I_2
20. FeS_2
21. KOH
22. $KClO_4$
23. SnO_2
24. C_6H_6
25. Hg_2Cl_2
26. HNO_3
27. $PbBr_2$
28. CaO
29. H_2SO_4
30. $Al(OH)_3$

Student Lab: Empirical Formula

Sometimes it is valuable for a chemist to know the amount of each element that has combined to make a material. The purpose of this lab is to show you how to determine the empirical formula of a compound. An **empirical formula** shows the simplest ratio in which atoms can combine to form a compound.

Materials

- Bunsen burner
- Magnesium ribbon
- Crucible
- Crucible cover
- Ring stand
- Ring
- Clay triangle
- Tongs
- Digital scales or lab balance

Safety Considerations

Students should wear goggles.

All burning should be done with proper ventilation or in a fume hood, and the smoke given off by burning magnesium should not be inhaled.

Students should avoid looking at the burning magnesium. It contains unsafe amounts of ultraviolet radiation for viewing.

The crucible and cover will get extremely hot in the Bunsen burner flame. Use only the tongs to move them throughout the lab.

Procedure

1. Place the ring on the ring stand.
2. Place the clay triangle on the ring.
3. Place the crucible on the clay triangle with the crucible cover on it.

4. Heat the crucible and clay for 5 minutes in the clear blue of the flame to avoid soot buildup.
5. Turn off the burner and let the crucible and cover cool for 10 minutes.
6. Move the crucible and cover to a balance with a pair of tongs and record their mass. Do not touch them at any point in the lab.
7. Coil a 10–15 cm piece of magnesium ribbon and place it in the crucible.
8. Record the mass of magnesium, crucible, and cover.
9. Place the crucible with the magnesium in it back on the clay triangle. Replace the cover but leave it at an angle to allow air to enter the crucible while it is being heated.
10. If the magnesium ignites do not look directly at it. The light can damage your eyes, and there is ultraviolet radiation being given off.
11. Heat in the clear, blue flame for 10 minutes and then remove the burner and check under the cover. If any of the ribbon appears unburned return the burner and heat for another 5 minutes, and repeat until the metal has all been reacted.
12. When the material is all a light gray powder, you have a sample of magnesium oxide.
13. Allow the crucible, cover, and magnesium oxide to cool for 10 minutes, in a desiccating jar if available.
14. After cooling, measure the mass of the crucible, cover, and magnesium oxide.

Data Table

Mass of crucible and cover (g) ____________

Mass of the crucible, cover, and magnesium (g) ____________

Mass of the crucible, cover, and magnesium oxide (g) ____________

Calculations

Show your work.

1. Calculate the mass of the magnesium ribbon.
2. Calculate the moles of magnesium ribbon used.

3. Calculate the mass of magnesium oxide that was formed.

4. Calculate the mass of oxygen that combined with the magnesium ribbon.

5. Calculate the moles of oxygen that were combined with the magnesium ribbon.

6. Calculate the ratio between moles of magnesium atoms used and moles of oxygen used.

7. Calculate the empirical formula for magnesium oxide indicated by your data.

Comprehension Questions

1. What would the empirical formula be for magnesium oxide in each of the ratios below?

   a. 1:1

   b. 1:1.5

   c. 1:2

   d. 1:2.5

2. Which of the above formulas most closely matches your result?

Solutions

When dealing with a solution, it becomes necessary to have a way of determining the amount of dissolved material that is present in it. A **solution** is a homogenous mixture of molecular substances, made up of a solute and a solvent. A **solvent** is that part of a solution in which the solute is dissolved, and a **solute** is that part of a solution that is the dissolved substance. When salt dissolves into water, water is the solvent and salt is the solute.

Molarity is the concentration of a solution given in moles of solute per liter of solution.

Knowing that we can dissolve different amounts of material into a solvent—say a teaspoon of salt into a gallon of water, or 20 teaspoons of salt into a gallon of water—requires us to differentiate between the concentrations of those two solutions. The **concentration** is the amount of solute in a given volume of solvent. **Molarity** is the concentration of a solution given in moles of solute per liter of solution. A 1 molar solution of NaCl in water would have 1 mole of salt in 1 liter of solution. A 3 molar solution of salt would have 3 moles of salt in one liter of solution.

Sometimes a solution is too concentrated to be used as it is and must be diluted. **Dilution** reduces the concentration of a solution, usually by adding more solvent. In most chemistry labs, acids and bases are ordered in concentrated form because they take up less room in shipping and storage; then they are added to some amount of water to dilute them into a concentration appropriate for everyday use.

One way to determine concentration is by titration. **Titration** determines the concentration of a solution by reacting it with a solution of known concentration. If two chemicals are known to react in a 1:1 ratio, then a solution of known concentration, often called a standard solution, can be reacted with a known chemical that has an unknown concentration. This process is usually carried out in the presence of a third chemical called an indicator, which will change the color of the mixing solutions when one or the other of the chemicals has been completely used up, or is in excess. If 1 milliliter of standard is used up in a reaction with 1 milliliter of unknown solution, then they have the same concentration. If twice as much known solution is used, then the concentration of the unknown is double that of the known solution.

Exploration Activities

1. What kind of substance can be used in titration to tell whether or not two solutions have completely reacted?

2. How many moles of salt are in a 2.5 molar solution if there is 1 liter of solution?

3. What would be the molarity of 12 moles of salt in 16 liters of solution?

The Periodic Table

One of the many ways scientists use to classify substances is the periodic table. The periodic table is a chart showing the elements in order by increasing atomic number and grouped by their similar qualities. German chemists Johann Döbereiner and Lother Meyer, as well as English chemist Jon Newlands, all made major contributions to the creation of the periodic table and found many ways of grouping and identifying the elements, particularly by each element's mass. However, Russian chemist Dimitri Mendeleev is believed to be the first to place the elements in order by their increasing atomic masses. There is some debate that says Meyer figured out the system first, but in their later years both Meyer and Mendeleev came to agree that they had both come up with the idea independently. Mendeleev realized that the repeating patterns he discovered as he listed each element were grouping the elements by their properties, a phenomenon called **periodicity;** he even went so far as to leave spaces for elements that had not yet been

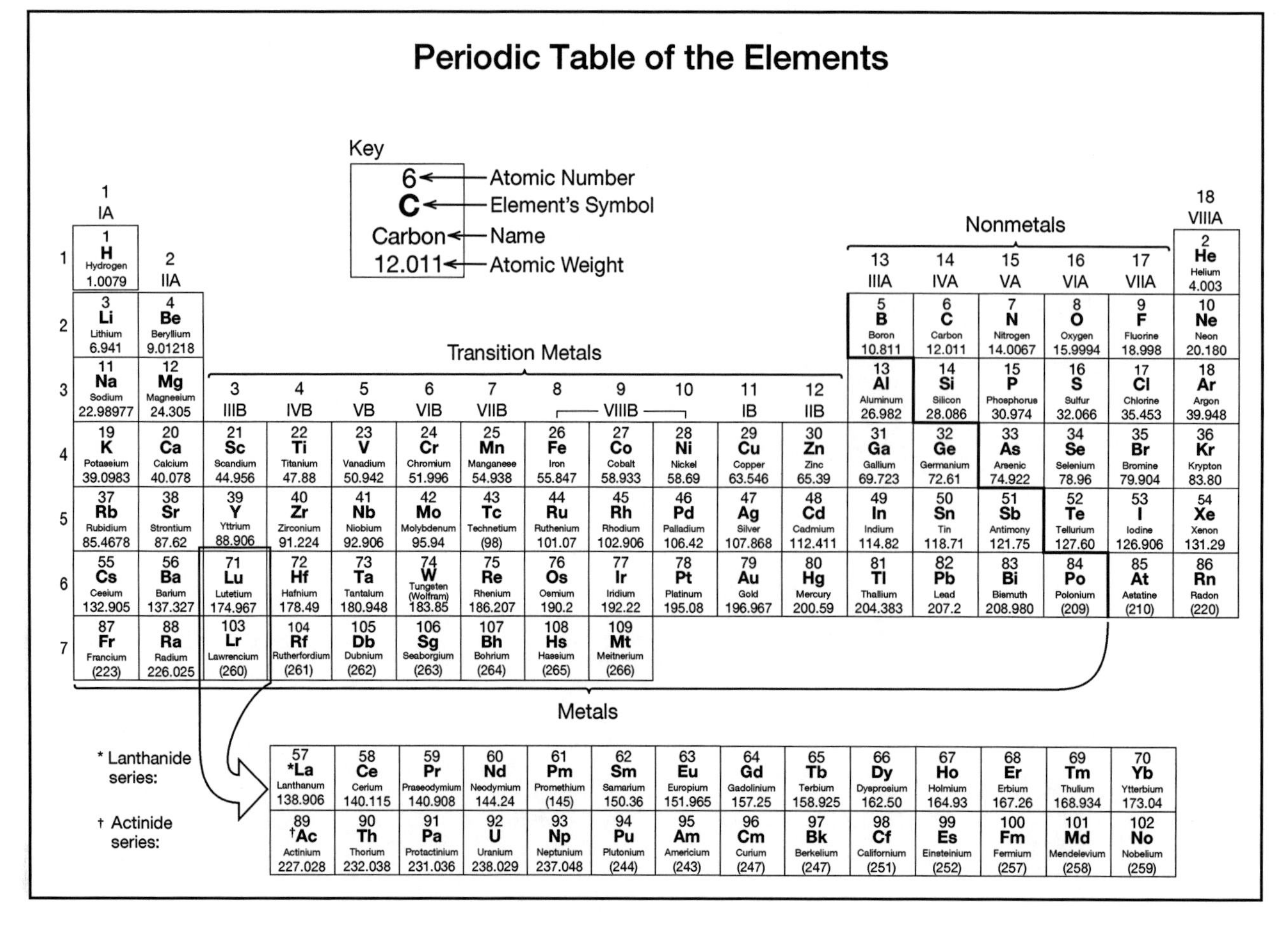

Periodic Table of the Elements

Key: 6 ← Atomic Number; C ← Element's Symbol; Carbon ← Name; 12.011 ← Atomic Weight

Metals (groups 1–12 and Al, Ga, In, Sn, Tl, Pb, Bi, Po); Transition Metals (groups 3–12); Nonmetals (groups 13–17)

Period	1 IA	2 IIA	3 IIIB	4 IVB	5 VB	6 VIB	7 VIIB	8 VIIIB	9 VIIIB	10 VIIIB	11 IB	12 IIB	13 IIIA	14 IVA	15 VA	16 VIA	17 VIIA	18 VIIIA
1	1 **H** Hydrogen 1.0079																	2 **He** Helium 4.003
2	3 **Li** Lithium 6.941	4 **Be** Beryllium 9.01218											5 **B** Boron 10.811	6 **C** Carbon 12.011	7 **N** Nitrogen 14.0067	8 **O** Oxygen 15.9994	9 **F** Fluorine 18.998	10 **Ne** Neon 20.180
3	11 **Na** Sodium 22.98977	12 **Mg** Magnesium 24.305											13 **Al** Aluminum 26.982	14 **Si** Silicon 28.086	15 **P** Phosphorus 30.974	16 **S** Sulfur 32.066	17 **Cl** Chlorine 35.453	18 **Ar** Argon 39.948
4	19 **K** Potassium 39.0983	20 **Ca** Calcium 40.078	21 **Sc** Scandium 44.956	22 **Ti** Titanium 47.88	23 **V** Vanadium 50.942	24 **Cr** Chromium 51.996	25 **Mn** Manganese 54.938	26 **Fe** Iron 55.847	27 **Co** Cobalt 58.933	28 **Ni** Nickel 58.69	29 **Cu** Copper 63.546	30 **Zn** Zinc 65.39	31 **Ga** Gallium 69.723	32 **Ge** Germanium 72.61	33 **As** Arsenic 74.922	34 **Se** Selenium 78.96	35 **Br** Bromine 79.904	36 **Kr** Krypton 83.80
5	37 **Rb** Rubidium 85.4678	38 **Sr** Strontium 87.62	39 **Y** Yttrium 88.906	40 **Zr** Zirconium 91.224	41 **Nb** Niobium 92.906	42 **Mo** Molybdenum 95.94	43 **Tc** Technetium (98)	44 **Ru** Ruthenium 101.07	45 **Rh** Rhodium 102.906	46 **Pd** Palladium 106.42	47 **Ag** Silver 107.868	48 **Cd** Cadmium 112.411	49 **In** Indium 114.82	50 **Sn** Tin 118.71	51 **Sb** Antimony 121.75	52 **Te** Tellurium 127.60	53 **I** Iodine 126.906	54 **Xe** Xenon 131.29
6	55 **Cs** Cesium 132.905	56 **Ba** Barium 137.327	71 **Lu** Lutetium 174.967	72 **Hf** Hafnium 178.49	73 **Ta** Tantalum 180.948	74 **W** Tungsten (Wolfram) 183.85	75 **Re** Rhenium 186.207	76 **Os** Osmium 190.2	77 **Ir** Iridium 192.22	78 **Pt** Platinum 195.08	79 **Au** Gold 196.967	80 **Hg** Mercury 200.59	81 **Tl** Thallium 204.383	82 **Pb** Lead 207.2	83 **Bi** Bismuth 208.980	84 **Po** Polonium (209)	85 **At** Astatine (210)	86 **Rn** Radon (220)
7	87 **Fr** Francium (223)	88 **Ra** Radium 226.025	103 **Lr** Lawrencium (260)	104 **Rf** Rutherfordium (261)	105 **Db** Dubnium (262)	106 **Sg** Seaborgium (263)	107 **Bh** Bohrium (264)	108 **Hs** Hassium (265)	109 **Mt** Meitnerium (266)									

Series														
* Lanthanide series:	57 ***La** Lanthanum 138.906	58 **Ce** Cerium 140.115	59 **Pr** Praseodymium 140.908	60 **Nd** Neodymium 144.24	61 **Pm** Promethium (145)	62 **Sm** Samarium 150.36	63 **Eu** Europium 151.965	64 **Gd** Gadolinium 157.25	65 **Tb** Terbium 158.925	66 **Dy** Dysprosium 162.50	67 **Ho** Holmium 164.93	68 **Er** Erbium 167.26	69 **Tm** Thulium 168.934	70 **Yb** Ytterbium 173.04
† Actinide series:	89 **†Ac** Actinium 227.028	90 **Th** Thorium 232.038	91 **Pa** Protactinium 231.036	92 **U** Uranium 238.029	93 **Np** Neptunium 237.048	94 **Pu** Plutonium (244)	95 **Am** Americium (243)	96 **Cm** Curium (247)	97 **Bk** Berkelium (247)	98 **Cf** Californium (251)	99 **Es** Einsteinium (252)	100 **Fm** Fermium (257)	101 **Md** Mendelevium (258)	102 **No** Nobelium (259)

discovered. He was proven out in his theory by the discovery of three elements, gallium, scandium, and germanium, and some scientists give him credit for predicting the location and properties of perhaps eight or more elements that were not known when he first proposed his theory.

In the periodic table, there are areas where elements having certain properties are found. For example, the metals are found toward the left and middle of the chart. A **metal** is an element that loses electrons easily in a chemical change and has the properties of high luster, electrical and thermal conductivity, malleability, and ductility. Some examples of metals are sodium, chromium, and copper.

Toward the top right are the nonmetals. A **nonmetal** is an element that easily gains electrons during a chemical reaction and whose properties are the opposite of those of the metals. Some nonmetals are oxygen, nitrogen, and chlorine.

Sandwiched between the metals and the nonmetals, on the staircase-like line drawn on many periodic tables from boron down to astatine, are the metalloids. A **metalloid (semimetal)** is an element that has both metallic and nonmetallic properties. Some metalloids are boron, silicon, and germanium.

EXPLORE

Exploration Activities

1. How are elements ordered on the modern periodic table?

2. Identify the following as metal, nonmetal, or metalloid.

   a. Silver

   b. Calcium

   c. Sulfur

   d. Antimony

3. Is a metal more likely to gain or lose electrons during a chemical reaction?

BACKGROUND

Types of Reactions

There are countless different ways that elements and compounds can interact with one another.

There are countless different ways that elements and compounds can interact with one another. One of the ways that scientists further identify the substances they are working with is to try to classify the various ways that they can interact. There are many kinds of reactions and combinations of reactions, but here are five basic reactions that are seen often in the study of chemistry. In the representations of each formula the letters A–D have been used to represent any possible element or compound, and then a real reaction has been included.

1. **Combination reaction (synthesis)**—a reaction in which two or more elements combine to form a single substance

   $$A + B \rightarrow AB$$
   $$2H_2 + O_2 \rightarrow 2H_2O$$

2. **Decomposition reaction (analysis)**—a reaction in which a single substance is separated into two separate substances

   $$AB \rightarrow A + B$$
   $$2H_2O \rightarrow 2H_2 + O_2$$

3. **Single-Replacement Reaction**—a reaction in which an element replaces a less chemically active element in a compound; the replaced element is set free.

   $$A + BC \rightarrow B + AC$$
   $$Cu + 2AgNO_3 \rightarrow 2Ag + Cu(NO_3)_2$$

4. **Double-replacement reaction**—a reaction in which two ionic compounds exchange ions to create two new ionic compounds

   $$AB + CD \rightarrow AD + CB$$
   $$AgNO_3 + NaCl \rightarrow AgCl + NaNO_3$$

5. **Combustion Reaction (Hydrocarbon Combustion)**—a reaction in which a hydrocarbon is burned and consumed in the

presence of oxygen. In the example below, C_xH_y is representative of any possible hydrocarbon; e.g., CH_4 or C_2H_6.

$$C_xH_y + O_2 \rightarrow CO_2 + H_2O$$
$$2C_2H_6 + 7O_2 \rightarrow 4CO_2 + 6H_2O$$

Displacement Reactions and the Activity Series

Activity Series (from most active to least active)

Li
K
Ba
Sr
Ca
Na
Mg
Al
Mn
Zn
Fe
Cd
Ca
Ni
Sn
Pb
H
Cu
Ag
Hg
Au

There is a kind of single-replacement reaction that is often called a displacement reaction. A displacement reaction is a single-replacement reaction in which a free metal replaces a metal in a compound. The free metal must have a higher chemical reactivity than the one in the compound, or the reaction will not take place.

The table at left shows a list of metals arranged from most active to least active. For example, the reaction below can take place

$$Li + AgNO_3 \rightarrow Ag + LiNO_3$$

because lithium is a more reactive element than silver. However,

$$Ag + LiNO_3 \rightarrow Li + AgNO_3$$

is a reaction that does not take place because silver doesn't have enough reactive energy to "push" lithium out of the compound.

Precipitation Reactions

Some reactions are carried out in solution to allow for the exchange of ions. In a **precipitation** reaction, one of the products of the reaction is a precipitate. A precipitate is an insoluble solid that is formed in a solution and settles out. There are many ways for precipitates to form, but usually they form during double-replacement reactions between strong **electrolytes.**

$$AgNO_3(aq) + NaCl(aq) \rightarrow NaNO_3(s) + AgCl(aq)$$

Those substances followed by an (*aq*) are the ones that are still in aqueous solution—that is, still dissolved in water. The (*s*) indicates that a solid was formed that cannot dissolve in water.

EXPLORE Exploration Activities

Identify each type of reaction.

1. $Pb(NO_3)_2 + KI \rightarrow PbI_2 + KNO_3$
2. $Al + Cl_2 \rightarrow Al_2Cl_3$
3. $C_2H_6 + O_2 \rightarrow CO_2 + H_2O$
4. $HCl + Sr(OH)_2 \rightarrow SrCl_2 + H_2O$
5. $Zn + HCl \rightarrow ZnCl_2 + H_2$
6. $K_2CO_3 \rightarrow K_2O + CO_2$
7. $SO_2 + O_2 \rightarrow SO_3$
8. $Fe_3O_4 + H_2 \rightarrow Fe + H_2O$
9. $AgNO_3 + MgCl_2 \rightarrow AgCl + Mg(NO_3)_2$
10. $KClO_3 \rightarrow KCl + O_2$
11. $CH_4 + Cl_2 \rightarrow CCl_4 + H_2$
12. $C_3H_8 + O_2 \rightarrow CO_2 + H_2O$
13. $O_2 + N_2 \rightarrow N_2O_4$
14. $NaF \rightarrow Na + F$
15. $NaI + Pb(NO_3)_2 \rightarrow NaNO_3 + PbI_2$
16. $Li_3PO_4 + Li \rightarrow P + O_2$
17. $CS_2 + F_2 \rightarrow CF_4 + S$

Identify the precipitate in each of the following reactions.

18. $Pb(NO_3)_2(aq) + K_2CrO_4(aq) \rightarrow PbCrO_4(s) + 2KNO_3(aq)$
19. $Ba(NO_3)_2(aq) + 2NH_4IO_3(aq) \rightarrow Ba(IO_3)_2(s) + 2NH_4NO_3(aq)$

Acid–Base Reactions

There is a large number of reactions that take place between acids and bases. Although there has been a lot of disagreement in the past about the exact properties that make up acids and bases, there are some general attributes they seem to have. An **acid** is (1) a substance that can be a proton donor, (2) a substance that can act as an electron acceptor, (3) a substance that produces hydronium ions (H_3O^{1+}) when dissolved in water. A **base** is (1) a substance that can be a proton acceptor, (2) a substance that can act as an electron donor, (3) a substance that produces hydroxyl ions (OH^{1-}) when dissolved in water. A proton is usually donated when a hydrogen atom is pulled away from a compound and leaves its only electron behind. The particle that remains is a H^+ ion, essentially the hydrogen's nucleus, which was only a proton to start with.

Many acids and bases will dissociate into ions when dissolved into water.

Many acids and bases will dissociate into ions when dissolved into water. For example, the reaction of hydrochloric acid, HCl, with water produces the following:

$$HCl + H_2O \rightarrow H_3O^{1+} + Cl^{1-}$$

while a reaction of a base, sodium hydroxide, NaOH, also produces ions.

$$NaOH + H_2O \rightarrow NH_4^{1+} + OH^{1-}$$

Often the reaction of an acid with a base will result in salt and water as the **products;** this is called a **neutralization reaction.** This is only true for a certain set of acids and bases, and two examples are shown below. Hydrochoric acid reacts with sodium hydroxide, a base, to form sodium chloride, a salt, and water.

$$HCl + NaOH \rightarrow NaCl + H_2O$$

In the second example sulfuric acid reacts with sodium hydroxide to form sodium sulfate, which is a salt, and water.

$$H_2SO_4 + 2NaOH \rightarrow Na_2SO_4 + 2H_2O$$

Water itself tends to be self-ionizing. That is, even just a sample of pure water has the ability to form the common ions found in acid and base reactions. Although the amount of ionization in pure water is small, it does exist and is represented by the formula below:

$$H_2O \rightleftarrows H^+ + OH^-$$

The left-pointing arrow indicates that the reaction tends to favor the existence of a lot more water than of separate ions.

Applying the law of chemical equilibrium we get:

$[H^+][OH^-] = K_w$

$[H_2O]$

where the ions in brackets represent the concentration of each substance in the solution, and K_w is the ion product constant for water. By experiment scientists have determined that the value of K_w, which is also $[H^+][OH^-]$, to be 1.0×10^{-14}. The current practice in chemistry is to evaluate the acidic or basic nature of a substance by calculating the pH of a solution. pH is found by taking the negative log, to the base 10, of the hydrogen ion.

$pH = -\log [H^+]$

and is sometimes given as

$pH = -\log [H_3O^+]$

When the value of this expression is higher than 7, the solution is basic. When the value is under 7, the solution is acidic, and a value of exactly 7 is considered neutral.

EXPLORE

Exploration Activities

1. Identify the acid, the base, and the salt in the following reaction.

$$H_2SO_4 + 2NaOH \rightarrow Na_2SO_4 + 2H_2O$$

2. Calculate the pH of a solution that has an H^+ concentration of

   a. 1.0×10^{-4} M

   b. 1.0×10^{-8} M

   c. 1.0×10^{-12} M

3. Calculate the original concentration of H_3O^+ ions in solutions with the following pHs.

   a. 7

   b. 9

   c. 6

   d. 2

Student Lab: Chemical Reaction Types

Although there are many kinds of reactions, there are five reactions common in introductory chemistry that are simple to identify. The goal of this lab is to familiarize you with basic reaction types, and teach you to identify them and write out chemical equations for those reactions.

Materials

- Disposable butane lighter
- Magnesium ribbon
- Two well plates per student or group
- Granular zinc
- Granular copper
- Granular aluminum
- 1 molar solution of the following in eyedroppers:
  $AgNO_3$
  $CuSO_4$
  HCl
  NaCl
  NaOH
  $Pb(NO_3)_2$

Safety Considerations

Goggles and lab aprons should be worn.

All chemicals should be considered toxic or corrosive.

All chemicals should be properly disposed of.

The Bunsen burner or butane lighter can make objects very hot. Handle with care.

Procedure

Part 1: Single-replacement reactions

Place the zinc, aluminum, and copper granules in the wells of the well plate as shown on the next page.

Add 10–20 drops of solution to each well, one at a time, as indicated below.

Record your observations about each reaction.

Use the activity series to predict whether or not each reaction is possible.

Well plate #1

Well 1: Zn(*s*)	+	$CuSO_4$(*aq*)	Well 5: Al(*s*)	+	$CuSO_4$(*aq*)
Well 2: Zn(*s*)	+	$AgNO_3$(*aq*)	Well 6: Cu(*s*)	+	$AgNO_3$(*aq*)
Well 3: Zn(*s*)	+	$Pb(NO_3)_2$(*aq*)	Well 7: Cu(*s*)	+	$Pb(NO_3)_2$(*aq*)
Well 4: Zn(*s*)	+	HCl(*aq*)	Well 8: Al(*s*)	+	HCl(*aq*)

Part 2: Double-replacement reactions

Use 10–15 drops of each chemical to fill the wells of the well plate as indicated below.

Observe each reaction carefully and record your observations.

Use the solubility chart to determine which of the following reactions will take place.

Well plate #2

Well 1: NaCl(*aq*)	+	$CuSO_4$(*aq*)
Well 2: NaOH(*aq*)	+	$CuSO_4$(*aq*)
Well 3: NaOH(*aq*)	+	$Pb(NO_3)_2$(*aq*)
Well 4: NaCl(*aq*)	+	$AgNO_3$(*aq*)
Well 5: NaCl(*aq*)	+	$Pb(NO_3)_2$(*aq*)
Well 6: NaOH(*aq*)	+	$AgNO_3$(*aq*)

Solubilities of Selected Compounds in Water

Compound	Solubility in Water	Compound	Solubility in Water
AgOH	Insoluble	NaCl	Soluble
AgCl	Insoluble	$NaNO_3$	Soluble
$AgNO_3$	Soluble	NaOH	Soluble
$Cu(OH)_2$	Insoluble	$Pb(NO_3)_2$	Soluble
$CuSO_4$	Soluble	$Pb(OH)_2$	Insoluble
Na_2SO_4	Soluble	$PbCl_2$	Insoluble

Part 3: Combustion—Teacher Demonstration

A disposable butane lighter is a common device used to make fire. Each time it is lit, it produces the combustion reaction shown below.

$$2C_4H_{10} + 13O_2 \rightarrow 8CO_2 + 10H_2O + \text{(Incomplete products like carbon soot)}$$

By holding a watch glass above the flame, you can capture the black carbon soot to show the incomplete combustion of the butane. You can also move the flame across the cool surface of a beaker to see the water vapor condense.

Part 4: Combination—Teacher Demonstration

Students may have already done the lab called Empirical Formula (on page 45) in which they combined magnesium with oxygen to make magnesium oxide. If they haven't, this would be a good place to burn a piece of magnesium ribbon. If it is weighed beforehand and all the ashes are captured and weighed after it is burned, then it will weigh slightly more, indicating that it has gained mass in the combination process.

$$2Mg + O_2 \rightarrow 2MgO$$

Part 5: Decomposition—Teacher Demonstration

The electrolysis of water is a common demonstration to show the decomposition of a substance, and if you have access to a Hoffman apparatus it is quite easy to do. However, if you don't, you can build your own inexpensive Hoffman apparatus out of common household goods and a little silicone sealant.

1. Poke two pencil-sized holes in the sides of two 2-liter soda bottles, near the bottom.
2. Put a ring of silicone sealant around the holes, and then tape the bottles together to make a waterproof window connecting them.
3. On opposite sides of the bottles, again near the bottom, jab two nails through the bottles and seal them in place with silicone.
4. Fill the two bottles with a teaspoon of white vinegar and water. Be sure that the bottles are completely filled with the water-vinegar solution as we want a minimum of air in the bottles.
5. Drill a hole in each bottle's plastic cap. Use silicone to seal a medicine dropper through each cap, forming a spout with the narrow end protruding from the top of each bottle.
6. Screw the caps on the bottles.

7. The nails will be the electrodes for the electrolysis. Connect one nail to the (+) pole of a 6-volt lantern battery. Connect the other nail to the (–) pole of the battery, and electrolysis should begin with hydrogen rising at one electrode and oxygen at the other. The gases will emerge from the medicine droppers unless you seal them with some plastic wrap.

8. If you attach hoses to the jets you can deliver the gases into inverted, water-filled bottles to collect a sample of hydrogen and oxygen. Be aware of the fact that both gases are explosive, and oxygen dangerously so. DO NOT RECOMBINE THE GASES. A small amount of hydrogen gas and oxygen gas together is extremely explosive and powerful.

9. Disconnect the battery when done, and dispose of any gas you do not use in the fume hood or outside.

Comprehension Questions

1. Make an organized list of all the observations that indicated a reaction was taking place in part 1. Be sure to indicate which well of the well plate you are referring to.

2. Make an organized list of all the observations that indicated a reaction was taking place in part 2. Be sure to indicate which well of the well plate you are referring to.

3. Write out chemical equations for all of the reactions in parts 1 and 2. Indicate whether the material was aqueous (*aq*), solid (*s*), or gaseous (*g*).

BACKGROUND

Oxidation Numbers

Oxidation number is the effective charge of an atom when it is chemically combined in a compound. In many cases, the oxidation number of an element is the same as the charge determined for an ion of that element. One difference, however, is the way the charges or oxidation numbers are written—the placement of the plus or minus sign that indicates whether the number is positive or negative. For example, the charge on the lithium ion is 1+, but the oxidation number is +1. It might seem like a minor difference, but it is necessary when working on a complicated matter like the oxidation numbers of elements that make up a compound or polyatomic ion.

The charge on the lithium ion is 1+, but the oxidation number is +1.

Some Guidelines for Oxidation Numbers

Some chemical equations in the following examples have been simplified to better show the oxidation numbers and ionic charges.

1. The oxidation number of any free element is zero. Carbon, in an uncombined state, has an oxidation number of zero. Any diatomic element like bromine, which usually appears in nature as Br_2, has an oxidation number of zero.

2. The **oxidation** of any single element in a compound is equal to the charge of that element as an ion in that compound. For example, in NaCl, the sodium ion is Na^{1+}, and the chloride ion is Cl^{1-}, whereas the oxidation number of sodium is +1, and the oxidation number of the chlorine is −1.

3. The sum of the oxidation numbers in a compound is zero, and the sum of the oxidation numbers in an ion is equal to the charge of the ion. For example, in the compound $MgCl_2$:

$$Mg^{2+} + 2Cl^{1-} \rightarrow MgCl_2$$
$$+2 + 2(-1) = 0$$

and in the polyatomic ion $NO_3{}^{1-}$:

$$N^{5+} + 3O^{2-} \rightarrow NO_3{}^{1-}$$
$$+5 + 3(-2) = -1$$

4. Hydrogen usually has an oxidation number of +1, unless it is combined with one of the group 1 or 2 metals to form a hydride, in which case its oxidation number is –1. For example:

$H^{1+} + Cl^{1-} \rightarrow HCl$

$Ba^{2+} + 2H^{1-} \rightarrow BaH_2$ (barium hydride)

5. Oxygen usually has an oxidation number of –2, unless it is in a peroxide, in which case its oxidation number is –1.

$Ca^{2+} + O^{2-} \rightarrow CaO$

$2H^{1+} + 2O^{2-} \rightarrow H_2O_2$

Exploration Activities

1. What is the relationship between the oxidation number of a free element and that same element as an ion?

2. What is the oxidation number for each element in the compound Al_2O_3?

3. Show that the sum of all the oxidation numbers in the ion SO_3^{2-} is –2.

Oxidation-Reduction Reactions

Oxidation is a reaction in which a particle, such as an ion, an atom, or a molecule, loses an electron. When this occurs, the oxidation number of some element in the reaction increases. For example, in the reaction below, the down arrows indicate the oxidation number of each element.

$$\begin{array}{ccccc} 2Na & + & Cl_2 & \rightarrow & 2NaCl \\ \downarrow & & \downarrow & & \downarrow \; \downarrow \\ 0 & + & 0 & \rightarrow & +1 \; -1 \end{array}$$

OIL RIG stands for Oxidation Is Losing (of electrons) and Reduction Is Gaining (of electrons).

As you can see, the oxidation number of sodium increased from zero to +1; this reaction also gives us an example of reduction. **Reduction** is a reaction in which a particle, such as an ion, an atom, or a molecule, gains an electron. In this case it was the chlorine atom that gained an electron, changing its oxidation number from zero to −1. A memory trick for remembering the difference between oxidation and reduction used by many chemists is OIL RIG. The letters stand for Oxidation Is Losing (of electrons) and Reduction Is Gaining (of electrons).

The reaction above is actually an **oxidation-reduction reaction,** which is a reaction in which both an oxidation reaction and a reduction reaction take place. Another example would be the reaction below.

$$\begin{array}{ccccc} C & + & O_2 & \rightarrow & CO_2 \\ \downarrow & & \downarrow & & \downarrow \; \downarrow \\ 0 & & 0 & & +4 \; -2 \end{array}$$

Exploration Activities

1. Define both oxidation and reduction reactions in terms of both oxidation numbers and transfer of electrons.

2. In the following equations, identify the element that is oxidized and the element that is reduced.

   a. $CaBr_2 + Pb(NO_3)_2 \rightarrow Ca(NO_3)_2 + PbBr_2$

   b. $SnCl_2 + 2FeCl_3 \rightarrow SnCl_4 + 2FeCl_2$

   c. $4HCl + MnO_2 \rightarrow 2H_2O + Cl_2 + MnCl_2$

Balancing Chemical Equations

In describing the reactions that take place between various substances, it is very important to keep track of the amount of each **reactant** that is used and each product that is created. The law of conservation of mass tells us that the amount of mass you start with should be equal to the amount of mass that is found at the end of a reaction. Look at the unbalanced equation below:

$$H_2 \quad + \quad O_2 \quad \rightarrow \quad H_2O$$

In chemical equations, symbols to the left and right of the arrow must be the same in value.

Although it is true that hydrogen and oxygen combine to form water, a closer look at the equation will show that it is not true. The arrow in a **chemical equation** could be compared to the equal sign seen in basic addition. The equal sign means that the symbols found to the left and right of the sign are the same in value. For this to be true, we add coefficients in front of substances to make the equation true.

$$2H_2 \quad + \quad O_2 \quad \rightarrow \quad 2H_2O$$

The coefficient in front of hydrogen on the left multiplies it by two, making for a total of four hydrogen atoms. The coefficient on the right, in front of the water, multiplies every element by two, making four hydrogens and two oxygens. There are now an equal number of hydrogens and oxygens on each side of the equation.

Some Basic Balancing Guidelines

1. Be sure you can write the word equation. "Hydrogen combines with oxygen to form water" is a simple example.

2. Write the correct chemical formula for each substance in the reaction. There are diatomic elements, ions, and numerous other substances that can be combined in a wide variety of ways. If all such factors are not taken into account, an equation may be written with formulas that are incorrect, and may be impossible to balance.

3. Determine if coefficients must be added to make the equation balance. It's always worth checking to see if the equation needs to be balanced before trying coefficients randomly.

4. Be methodical. There are some rules for balancing various kinds of equations, but the best method to use is practice. Using trial and error, checking your answer, and checking to see that coefficients are the smallest whole numbers are all methods that will help you balance equations.

Now let's balance the reaction of aluminum chromate with barium chloride to form aluminum chloride and barium chromate.

$$Al_2(CrO_4)_3 + BaCl_2 \rightarrow AlCl_3 + BaCrO_4$$

Let's start by balancing the aluminum. There are two on the left, so we'll place a two in front of $AlCl_3$ on the right.

$$Al_2(CrO_4)_3 + BaCl_2 \rightarrow 2AlCl_3 + BaCrO_4$$

This has made six chlorines on the right, so let's put a three in front of $BaCl_2$ on the left.

$$Al_2(CrO_4)_3 + 3BaCl_2 \rightarrow 2AlCl_3 + BaCrO_4$$

This has made three bariums on the left, so we'll put a three in front of $BaCrO_4$ on the right.

$$Al_2(CrO_4)_3 + 3BaCl_2 \rightarrow 2AlCl_3 + 3BaCrO_4 \quad \text{(Balanced)}$$

Exploration Activities

1. Why is it necessary to balance chemical equations?

2. Write and balance the equation for calcium sulfate reacting with sodium carbonate to form calcium carbonate and sodium sulfate.

3. Write and balance the equation for zinc sulfide reacting with oxygen to form zinc oxide and sulfur dioxide.

Skill Builder: Balancing Chemical Equations

Balance the following equations.

1. $Na + H_2O \rightarrow NaOH + H_2$
2. $CaCO_3 + HCl \rightarrow CO_2 + CaCl_2 + H_2O$
3. $N_2 + H_2 \rightarrow NH_3$
4. $Ca(OH)_2 + H_2SO_4 \rightarrow CaSO_4 + H_2O$
5. $C_2H_2 + O_2 \rightarrow CO_2 + H_2O$
6. $NaCl \rightarrow Cl_2 + Na$
7. $Ti + N_2 \rightarrow Ti_3N_4$
8. $Al + ZnCl_2 \rightarrow AlCl_3 + Zn$
9. $Al + O_2 \rightarrow Al_2O_3$
10. $Fe + O_2 \rightarrow Fe_2O_3$
11. $CuO \rightarrow Cu + O_2$
12. $NO \rightarrow N_2 + O_2$
13. $NaCl + Li \rightarrow LiCl + Na$
14. $NaI + Br_2 \rightarrow NaBr + I_2$
15. $CaCO_3 + NH_4Cl \rightarrow (NH_4)_2CO_3 + CaCl_2$
16. $KOH + H_2SO_4 \rightarrow K_2SO_4 + H_2O$
17. $Al + H_2SO_4 \rightarrow Al_2(SO_4)_3 + H_2$
18. $Ca + H_3PO_4 \rightarrow Ca_3(PO_4)_2 + H_2$

19. $Mg + HNO_3 \rightarrow Mg(NO_3)_2 + H_2$

20. $C_{25}H_{52} + O_2 \rightarrow CO_2 + H_2O$

21. $Zn + HNO_3 \rightarrow Zn(NO_3)_2 + NH_4NO_3 + H_2O$

22. $NH_3 + NO \rightarrow N_2 + H_2O$

23. $C + SO_2 \rightarrow CS_2 + CO$

24. $Sb_2S_3 + Fe \rightarrow Sb + FeS$

25. $CaCO_3 + H_3PO_4 \rightarrow Ca_3(PO_4)_2 + CO_2 + H_2O$

Structure of the Atom

All of the interactions that take place in chemical reactions happen at very small levels. Humans have discovered numerous facts about the makeup of atoms and the ways in which they interact with one another. There are basic parts to the atom, and those parts, in turn, are made up of basic parts as well.

There are basic parts to the atom, and those parts, in turn, are made up of basic parts as well.

The electron is one of the three major subatomic particles. It has a negative **charge**, is found outside of the nucleus of the atom, and is symbolized as e^-. The first experimental evidence that led to the discovery of the electron is generally attributed to the work of English physicist William Crookes in the 1870s. He determined that some sort of ray or particle was produced in a vacuum tube when high voltage was passed through it. His work led to further experiments by English scientist J. J. Thomson, who discovered that the rays or particles coming off the cathode (the negatively charged electrode) could be deflected toward a positively charged metal plate. He concluded that these were negatively charged particles he called cathode ray particles. We now call these "electrons," and Thomson is usually thought of as their discoverer.

The proton is the second of the three major subatomic particles. It has a positive charge, is found in the nucleus, and has a mass similar to that of the neutron. The research that is generally considered to have led to the discovery of the proton was done by New Zealand scientist Ernest Rutherford and his two students, Hans Geiger and Ernest Marsden. They were working with a piece of machinery designed to shoot alpha particles at a piece of gold foil only a few atoms thick. They were fully expecting the vast majority of particles to pass through unaffected, but some of them bounced off at 90°, or almost straight back in the direction they had come from. Rutherford compared it to shooting an artillery shell at a piece of tissue paper and having it bounce off. The material they were hitting turned out to be the densely packed, positively charged nucleus of the atom.

The neutron is the third of the three major subatomic particles. It is electrically neutral and is found in the nucleus of the atom; it is similar in mass to the proton. James Chadwick built on the work of Irène Joliot-Curie, one of Madame Curie's daughters, and her husband, Frédéric Joliot-Curie. They had observed collisions in the lab and had incorrectly identified them as an effect that was not actually possible under the conditions they were testing. Chadwick bombarded beryllium atoms with alpha particles (helium nuclei) which produced carbon-12 atoms. He determined that there was a neutral component in the nucleus that had essentially the same mass as a proton. He called it the neutron.

Properties of the Three Major Subatomic Particles

Particle	Symbol	Charge	Mass in Grams
Electron	e^-	−1	9.109×10^{-28}
Proton	P	+1	1.673×10^{-24}
Neutron	N	0	1.675×10^{-24}

The nature of the particles found in an atom is quite wide and varied. There are groups of particles called fundamental particles, which are the leptons, bosons, and baryons. The leptons are a group of particles that includes electrons, muons, and tau particles. The bosons are photons, gluons, weakons, and possibly gravitons. And a third group is the baryons, which include the proton, neutron, hyperon, and the lambda, sigma, xi, and omega particles. There are many other particles that could be listed, but it is hard to give them day-to-day meaning considering the complicated ways in which they are generated and defined in physics. Suffice it to say that, as more and more of these particles are discovered, scientists come closer to understanding the fundamental forces of nature that cause matter to behave the way it does.

Already, many of the discoveries of the subatomic particles have led to a better understanding of the behavior of matter. British scientist H. G. Mosely was the first person to correctly determine atomic numbers. The **atomic number** is the number of protons in the nucleus of an atom. The atomic number is given as Z and has been determined so that each element has its own specific atomic number. For example, chlorine has an atomic number of $Z = 17$, so we know that it has 17 protons in its nucleus. Mercury has an atomic number of $Z = 80$, so we know that it has 80 protons in its nucleus. If an atom had 81 protons in its nucleus, it could not be mercury by definition.

The modern chemist makes much use of the process called spectroscopy. Spectroscopy is the technique of studying the light and other emissions of substances. Most of those emissions take the form of electromagnetic radiation. **Electromagnetic radiation** is an oscillating wave having both electric and magnetic components; it includes parts of the electromagnetic spectrum such as light and X rays. Different substances absorb or emit light based on their atomic or molecular structure. Each element has a unique structure that is revealed through spectroscopy. This allows scientists to take a complex sample, expose it to heat, and look at the spectrum of emissions. The emissions can then be compared to emissions of known elements and the elements that are present can be identified.

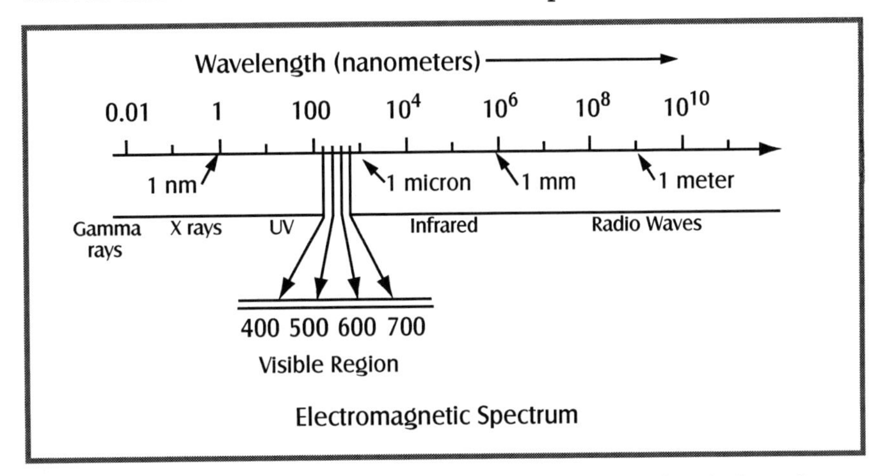

Electromagnetic Spectrum

Some substances eject photons when they are bombarded with energy. This is called the **photoelectric effect.** The photoelectric effect is the result of bombarding the surface of some metals with electromagnetic radiation so that the metal emits electrons. A common use for this effect is in any sort of electronic device that has a photoelectric cell, such as a solar-powered calculator. German physicist Max Planck was the first to think of these emissions as compact bundles of energy that he called quanta. He went on to clarify that the frequency of the radiation and the energy of the emitted photon were directly related.

Bohr Model of the Atom

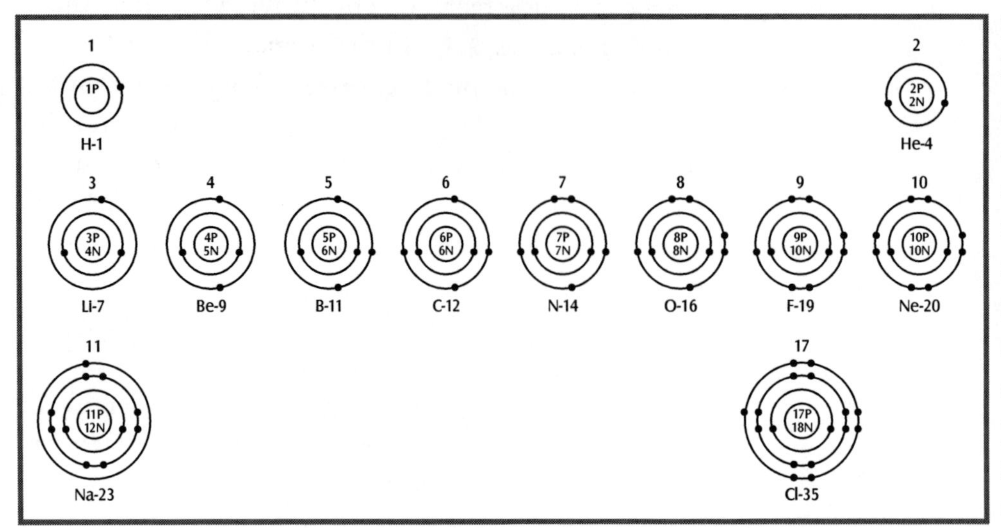

Danish physicist Niels Bohr made improvements to earlier models of the atom. He envisioned the atom as having a dense center that contained the protons and neutrons. The primary point of his model was that the electrons were in fixed orbits around the nucleus, and that every electron in an orbit had its own specific energy. More specifically, the orbits that were close to the nucleus had low energy, and as the orbits got farther away from the nucleus, the more energy they had. Bohr proposed that each atom had a ground state, at which electrons were in certain orbits and could only change orbit if they gained energy. If they did gain energy, they would move up to a higher energy level. If they lost energy they would drop back to the ground state, and the energy would be released as a photon of radiation. There is a limit to how many electrons can get into each level. There can be, at most, 2 electrons in the first level, 8 in the second, 18 in the third, 32 in the fourth, and, theoretically, 50 in the fifth; however, due to the order in which the energy levels fill, no naturally occurring element has a full fifth energy level. There are seven energy levels with electrons in them in the largest atoms, but none of the fifth, sixth, or seventh energy levels are filled in any atom.

Wave Nature of the Electron

While Max Planck was still working with the idea that it is possible for light to act as though it has the properties of particles, French physicist Louis de Broglie was proposing that waves could behave like a particle. In an experiment called the Davisson-Germer experiment, it was shown that electrons scattered from a metal surface as the surface was rotated through different angles.

It was shown that the energy of the scattered electrons could be increased and decreased in intensity according to the angle. The pattern also repeated, showing that it was wavelike in nature.

The Quantum Mechanical Picture of the Atom

The quantum mechanical view of atoms was first clearly stated by Austrian scientist Erwin Schrödinger in 1926. He formulated an equation that when solved gives a set of numbers, called quantum numbers, that describes the energies of electrons in atoms. Scientists use quantum numbers to describe the electron arrangement inside an atom, called the electron configuration. There are four main numbers that describe the location of electrons in the atom. The principal quantum number **n** describes the main energy level an electron occupies. The subsidiary quantum number **l** specifies sublevels within the main energy levels. This is also called the azimuthal number. The magnetic quantum number **m** describes the spatial orientation of an atomic orbital within a sublevel. The spin quantum number **s** refers to the spin of an electron, which is how it rotates as it moves.

Exploration Activities

1. Define the following terms:

   a. Leptons

   b. Atomic number

   c. Photoelectric effect

2. How were protons discovered?

3. How do the masses of the proton, neutron, and electron compare?

4. Draw a Bohr model for the elements hydrogen, chlorine, and argon.

5. As the wavelengths of the different kinds of electromagnetic radiation increase, their frequencies decrease. Which has a higher frequency, blue light or red light?

Student Lab: Atomic Structure and the Flame Test

When solutions of metals are heated in a Bunsen burner flame, they give off very specific colors. Neils Bohr proposed that there are regions in the atom where an electron naturally settles, and when that atom is energized—for example, with heat from a Bunsen burner—the electron jumps to a higher energy level, and then falls back to its original place. When this happens, the energy released as it falls is released in the form of a very specific frequency of light. For example, the copper in copper sulfate makes a green color with just a hint of blue when it is burned in a Bunsen burner, a color that is often seen in fireworks. If you have a wide variety of colors with which to compare, you can determine the makeup of an unknown substance by comparing it to the known colors.

Materials

- Test tube rack
- 10 flame test wires (nichrome)
- 5 molar hydrochloric acid
- Molar solutions of the following:
  Barium chloride
  Sodium chloride
  Calcium chloride
  Copper sulfate pentahydrate
  Lithium nitrate
  Potassium nitrate
  Three unknown solutions labeled 1, 2, and 3

Safety Considerations

Students should wear goggles and lab aprons.

Caution should be used with the Bunsen burners and anything that is placed into the flame—they will be extremely hot.

All chemicals should be treated as if toxic and corrosive.

Procedure

1. Get a clean flame test wire, and hold the metal loop in the hottest part of the Bunsen burner flame. If it is clean, there should be no change in the color of the flame. If it needs to be cleaned, clean it by dipping it into the concentrated acid provided and holding the loop in the Bunsen burner flame. Repeat if necessary until there is no change in the color of the flame.

2. Dip the clean flame test loop into one of the known test solutions; then hold the metal loop in the hottest part of the Bunsen burner flame. Carefully record the color of the flame. Try to make a comparison to a common item if you aren't sure that you will remember what the color looks like. For example, it might be the same color yellow as a common #2 pencil.

3. Clean the flame test wire and repeat the flame test with each of the known solutions.

4. Complete the flame tests for the three unknown solutions and make a note of their colors.

Comprehension Questions

1. What were the three unknown solutions? Identify each by number and by the known solution it compared most closely with.

2. What problems might arise if you needed to identify 300 chemicals by color alone?

3. What are some other ways that could be combined with this method that might make it easier to identify many different substances?

Electron Configurations

An **electron configuration** is the arrangement of electrons in the various orbitals of an atom. Electron configurations can be written using a form of notation that includes the principal energy level, the sublevel, and the number of electrons in that sublevel. For example, $3p^6$ means that this atom has electrons in the third principal energy level, in the *p* sublevel, and that there are six electrons. The sublevels are represented by the letters *s, p, d,* and *f,* which can contain at their maximum 2, 6, 10, and 14 electrons, respectively. They may have any number less than their maximum, or they may have their maximum, but any electrons beyond their maximum will usually go into the next available sublevel. Below is a periodic table that shows the locations of the energy levels and the sublevels.

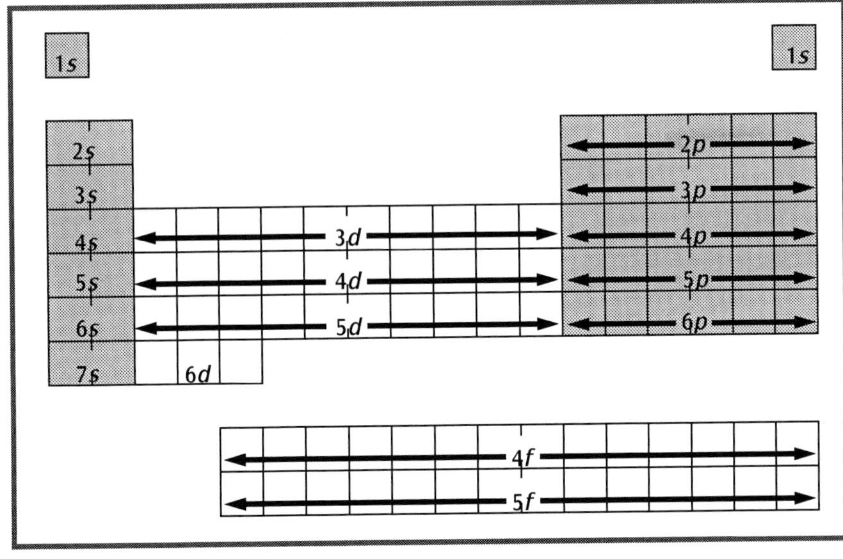

For example, hydrogen has an atomic number of 1, which means it has 1 electron. Its electron configuration is $1s^1$. Helium is number two and has a configuration of $1s^2$. As the number of electrons increases, the length of the electron configuration will also increase. The general order that the orbitals fill for the ground state of all the elements is

1s 2s 2p 3s 3p 4s 3d 4p 5s 4d 5p 6s 4f 5d 6p 7s 5f 6d 7p

Unfortunately, this is only a rough guideline, and there are exceptions. However, it is a good general guide. For example, cesium, $Z = 55$, has an electron configuration of $1s^2\ 2s^2\ 2p^6\ 3s^2\ 3p^6\ 4s^2\ 3d^{10}\ 4p^6\ 5s^2\ 4d^{10}\ 5p^6\ 6s^1$, which follows the guideline perfectly. A shortcut method of writing out the electron configurations is to use the noble gases as a starting point and then write only what would come after that noble gas. For example, xenon is number 54, so instead of writing out the entire configuration of cesium number 55 as above, you could write [Xe] $6s^1$.

Valence electrons are the electrons predominantly involved in the bonding process.

Valence Electrons

Valence electrons are the electrons in the outermost energy level of an atom, also called the valence shell. These are the electrons predominantly involved in the bonding process. For cesium the electron configuration was $1s^2\ 2s^2\ 2p^6\ 3s^2\ 3p^6\ 4s^2\ 3d^{10}\ 4p^6\ 5s^2\ 4d^{10}\ 5p^6\ 6s^1$, and by looking at the coefficients you can see that the highest energy level is 6. The exponent on the *s* is one, so the number of valence electrons for cesium is one. In an element like oxygen the electron configuration is $1s^2\ 2s^2\ 2p^4$, where the highest energy level is 2, but there are *s* electrons and *p* electrons and they must be added together to find the total number of valence electrons, which is $2 + 4 = 6$.

In a more complex configuration, like bromine, which is $1s^2\ 2s^2\ 2p^6\ 3s^2\ 3p^6\ 4s^2\ 3d^{10}\ 4p^5$, the highest energy level is 4, and even though the *s* and *p* electrons are separated by the *d* electrons in the third energy level, they are still both counted to find the total number of valence electrons, which is $2 + 5 = 7$.

Ionic Bonding

Ionic bonds are formed by the electrical attraction between positively charged **cations** and negatively charged **anions.** The attraction between Ca^{2+} cations and O^{2-} anions allows the compound CaO, calcium oxide, also known as lime, to be formed. The atoms involved in the bonding allow a transfer of electrons to take place, which creates a charge on each atom, making it an ion. The opposite charges then attract each other, forming a bond. In a strict sense, no bond is completely ionic because the electrons are not completely removed from the potential cation, but for most of the metals in the first two columns of the periodic table, the approximation is quite close. Generally, elements from groups 1 and 2 will lose electrons to reach a more stable internal arrangement, and elements from groups 15, 16, and 17 are likely to gain electrons for the same reason.

Lewis symbols, sometimes called dot diagrams, are ways of writing elemental symbols with dots around them (and sometimes small circles or the letter x) to indicate the electrons that are free to be

Some Lewis Symbols

Element	Electron Dot Symbol
Li	Li
Be	Be
B	B
C	C
N	N
O	O
F	F
Ne	Ne

involved in bonding. Elements from the first column of the periodic table are usually shown with one dot; column two gets two dots, column 13 gets three dots, column 14 gets four, column 15 gets five, column 16 gets six, and column 17 gets seven. Column 18, the noble gases, gets eight to show that their outer electron level is full. To the left is a chart showing the dot diagrams for period two of the periodic table.

When an ionic bond is formed between potassium and chlorine, the potassium loses one electron to the chlorine. This gives chlorine eight outer electrons, which simulates the behavior of the noble gases. The noble gases, except helium, have eight outer electrons and tend to neither gain nor lose electrons. Below is an example of an ionic bond being formed.

$K + Cl \rightarrow [K]^+[Cl]^-$

Or in the case of multiple ions being formed

$Ba + F + F \rightarrow [Ba]^{2+} \quad [F]^- \quad [F]^-$

Notice how the electron or electrons from the metal were transferred to fill the outer level of the nonmetal in each case.

Covalent Bonding

In ionic bonding, it is the transfer of electrons that makes up the bond that holds substances together, but in a **covalent bond** it is the sharing of electrons that gives bonds their strength. Atoms tend to form bonds to give them a more stable internal arrangement. In ionic bonding we saw that fluorine was able to get eight electrons into its outer level by gaining some electrons from barium. Typically, in covalent bonding a nonmetal bonds to itself or another nonmetal through the sharing of electrons. Notice that chlorine has seven outer electrons and needs to have eight to be the most stable. If two chlorines share one pair of electrons, both get to be surrounded by eight electrons.

$Cl + Cl \rightarrow Cl:Cl$

There are three common ways we symbolize such a bond.

Structural Formula	Molecular Formula	Dot Diagrams
Cl–Cl	Cl_2	$:\ddot{\underset{\cdot\cdot}{Cl}}:\ddot{\underset{\cdot\cdot}{Cl}}:$

Benzene

In the structural formula, the dash represents a shared pair of electrons. More complicated structural formulas are called Kekulé structures after German chemist Friedrich Kekulé, who proposed a simple notation system to show the structure of benzene.

Dipoles

A **dipole** is a molecule that has positive and negative locations within it despite the fact that the molecule as a whole is electrically neutral. The dipole typically has some positive charge at one end and some negative charge at the other end. Like the opposite north and south poles on the earth, which are magnetically opposite, a molecule can have ends that are electrically opposite. This arrangement in a molecule is sometimes referred to as *polar*. In the water molecule at left, you can see that, after the bonds are formed, there are some unpaired electrons left on one end of the water molecule.

This concentration of negative charges creates imbalance in the distribution of charge throughout the entire molecule. While the oxygen end has become slightly negative, the end with the two hydrogens has become slightly positive. The lowercase Greek letter sigma (δ) is used in this instance to indicate a partial charge on a polar molecule, and in general is used to mean "a small amount."

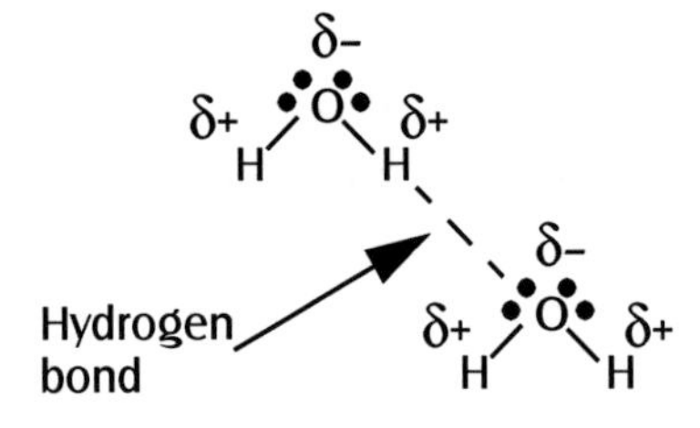

This arrangement in water allows water to create hydrogen bonds. **Hydrogen bonds** are very weak bonds that can be formed between a polar molecule with hydrogen and another molecule that is polar, like another water molecule. These weak bonds allow water molecules to be attracted to one another a little more than nonpolar molecules, and they account for some of water's unusual properties, like high surface tension and relatively high melting and boiling points.

Valence Shell Electron Pair Repulsion Theory

Valence shell electron pair repulsion (VSEPR) theory is a system for predicting the shape of molecules. The general shape of a molecule is determined by identifying the number of bonding and non-bonding electron pairs around a central atom, and then arranging them so that the repulsion electrons have for each other helps separate each electron pair or bonded atom so that they are as far apart as possible. The illustration below shows the variety of shapes and their names, plus some examples of molecules. See Appendix II for a chart of the various shapes and a list of the combinations of bonding electrons, nonbonding electrons, and atoms and the shapes they are likely to produce.

Number of Electron Pairs	Shape	Example	Number of Electron Pairs	Shape	Example
2	Linear 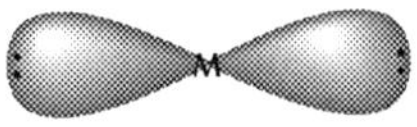	$BeCl_2$ 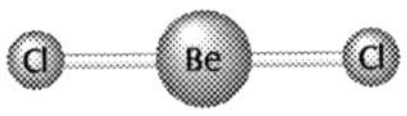	4	Tetrahedral 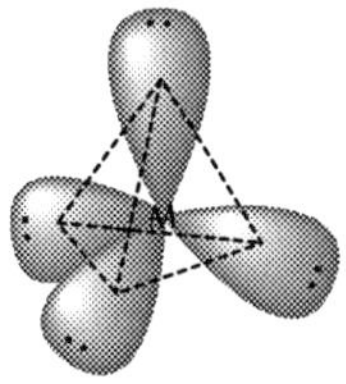	CH_4 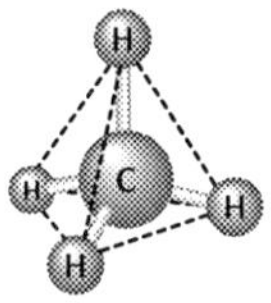
2	Bent 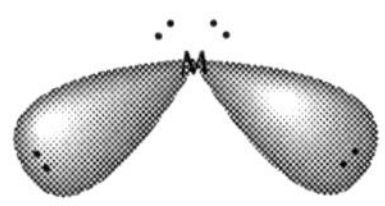	H_2O 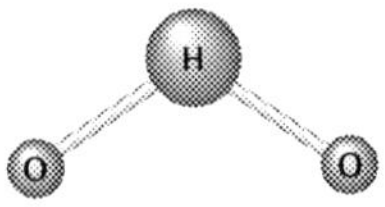	5	Trigonal bipyramidal 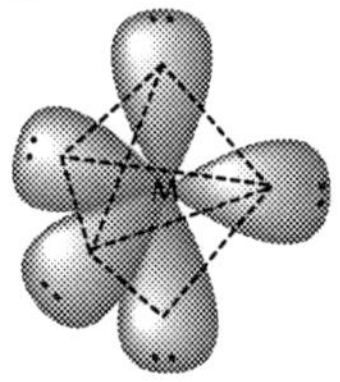	PCl_5 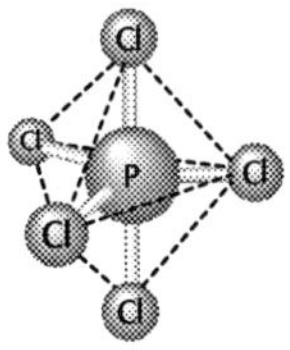
3	Trigonal Planar 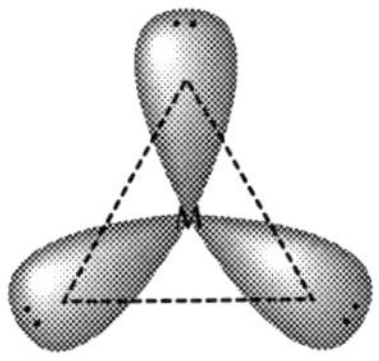	BCl_3 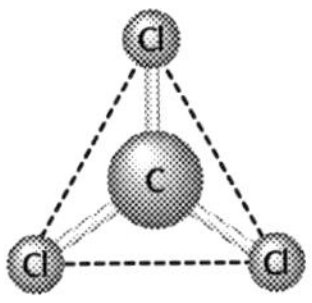	6	Octahedral 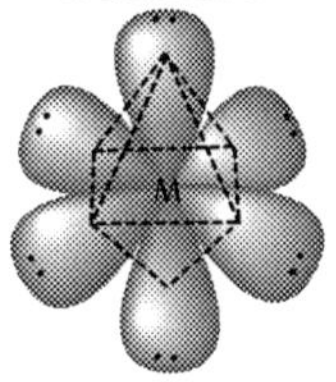	SF_6 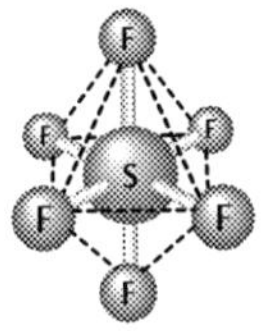

Exploration Activities

1. Write the electron configurations for each of the following atoms.
   a. Lithium
   b. Magnesium
   c. Argon
   d. Iodine
   e. Rubidium
2. How many valence electrons are there in each of the following atoms?
   a. Lithium
   b. Magnesium
   c. Argon
   d. Iodine
   e. Rubidium
3. Draw dot diagrams for each of the following.
   a. Lithium
   b. Magnesium
   c. Argon
   d. Iodine
   e. Rubidium
4. Predict the shapes of the following molecules.
   a. $NaCl$
   b. BeH_2
   c. SO_2
   d. SO_3
   e. H_2O
   f. NH_3
   g. CCl_4

Student Lab: Chemical Bonding

Atoms combine in many different ways, making for a wide variety of compounds and substances. One method of determining the chemical makeup of a substance after it has bonded is to compare the mass of the compound with the atomic masses of the reactants, and calculate what combination of original components would be needed to create a substance with the molecular mass of the product. This lab will show you how to use mass as a way of determining the ratios in which two substances can combine.

Materials

- Paper bags
- Nuts
- Bolts
- Digital scales or lab balance

Make up five paper bags by placing a random number of nuts and bolts into each bag. Label each bag, and keep track of what is in each one on a separate sheet of paper.

Safety Considerations

Although there is no chemical danger associated with this lab, students should wear goggles and be aware of the fact that nuts and bolts are not to be thrown for any reason.

Procedure

1. Mass an empty paper bag.
2. Mass 10 nuts.
3. Mass 10 bolts.
4. Mass unknown bag 1 and record your results.
5. Mass unknown bag 2 and record your results.
6. Mass unknown bag 3 and record your results.

7. Mass unknown bag 4 and record your results.
8. Mass unknown bag 5 and record your results.

Calculations

1. Calculate the mass of the material in bags 1–5 by subtracting out the mass of the paper bag.
2. Calculate the average mass of a single nut.
3. Calculate the average mass of a single bolt.
4. Determine the combination of nuts and bolts in each bag that would explain the mass of the material in the bag. Record your answer in the form Bo_XNu_Y where Bo is the chemical symbol for bolts, Nu is the chemical symbol for nuts, and *X* and *Y* are the number of bolts and nuts, respectively.

Comprehension Questions

1. Why would you want to do your calculations with the average mass of a nut and a bolt?

2. How does the concept of average mass compare to atomic mass and isotopes?

3. Calculate the mass of Nu_6Bo_{11} and show your work.

Gas Laws

Kinetic theory states that gases are made of small, constantly moving particles.

The behavior of gases is an area of chemistry about which very little was understood until the formulation of the kinetic theory. That theory gave scientists the idea that gases were made of small, constantly moving particles and allowed them to describe the behaviors they were seeing in terms of the actions of those particles. One of the things those particles do is collide with one another and with the surface of an object they come into contact with. These collisions exert a force on the surfaces with which they collide, and the amount of force they exert on a certain area of surface is known as pressure. **Pressure** is the amount of force exerted on a given area. In the SI system, the unit of pressure is the pascal (Pa), which is a N/m^2. The previous standard, and one that is still widely used, is the atmosphere. One atmosphere (atm) of pressure is the amount of pressure found at sea level, which is 101.325 kilopascals (kPa).

Robert Boyle was an Irish scientist who first expressed the relationship between the pressure and the volume of a gas. **Boyle's law** is the statement that the volume of a gas is inversely proportional to its pressure at a constant temperature. The mathematical relationship is then

$PV = K$ (at a constant temperature)
where
P = pressure
V = volume
K = a constant

So, in a closed system the product of the pressure and the volume should always be the same number, assuming the temperature is held constant. Therefore, if the initial pressure and volume of a system is known, and either the pressure or volume is changed, the product of the new values should equal the product of the original values, as expressed in the equation on the next page. (In all of the gas law descriptions, the word "initial" refers to the given values of the variables, while the word "final" refers to the value of the variables after the conditions have been changed.)

$P_1V_1 = P_2V_2$
where
P_1 = initial pressure
V_1 = initial volume
P_2 = final pressure
V_2 = final volume

Example 1

Suppose that 200 cm^3 of oxygen gas under 250 pascals is held at a constant temperature. What is the final volume of the gas if the pressure is increased to 500 pascals?

Solution:

$P_1V_1 = P_2V_2$
(250 Pa)(200 cm^3) = (500 Pa)(V_2)
V_2 = 100 cm^3

French scientist Jacques Charles had an interest in ballooning and how the behavior of gases might help him improve his ballooning skills. As an outgrowth of this interest, he was the first to express the relationship between temperature and the volume of a gas. **Charles' law** is the statement that the volume of a gas is directly proportional to its temperature at a constant pressure.

$V/T = K$ (at a constant pressure)
where
T = temperature (in Kelvin)
V = volume
K = a constant

So, in a closed system where the pressure is held constant, the ratio of volume divided by temperature will always be the same. Therefore, if the initial volume and temperature are known, and either one is changed, then it should be possible to calculate how the other is changed. The ratio of the new values should be equal to the ratio of the original values, as shown in the equation below.

$V_1/T_1 = V_2/T_2$
where
V_1 = initial volume
T_1 = initial temperature (in Kelvin)
V_2 = final volume
T_2 = final temperature (in Kelvin)

Example 2

500 liters of hydrogen gas at 300 K are cooled to 100 K while pressure is held constant. What is the final volume of gas?

$V_1/T_1 = V_2/T_2$
(500 L)/(300 K) = V_2/(100 K)
$V_2 = 167$ L

French scientist Joseph-Louis Gay-Lussac was a successful balloon pilot in the 1800s and set an altitude record of about 23,000 feet in 1804. His interest in ballooning led to his research into the behavior of gases, particularly how pressure and temperature are connected. **Gay-Lussac's law** is the statement that the pressure of a gas varies directly with the Kelvin temperature at a constant volume.

$P/T = K$ (at a constant volume)
where
T = temperature
P = pressure
K = a constant

So, in a closed system where the volume is held constant, the ratio of pressure divided by temperature will always be the same. Therefore, if the initial pressure and temperature are known, and either one is changed, then it should be possible to calculate how the other is changed. The ratio of the new values should be equal to the ratio of the original values, as shown in the equation below.

$P_1/T_1 = P_2/T_2$
where
P_1 = initial pressure
T_1 = initial temperature (in Kelvin)
P_2 = final pressure
T_2 = final temperature (in Kelvin)

Example 3

If you have a fixed volume of gas under 4 atm of pressure at 200°C, and the volume is changed to 20 atm, what is the final temperature of the gas? (Remember, all temperatures must be in Kelvin.)

Step 1: Convert 200°C to Kelvin: 200°C + 273 = 473 K

Step 2: Substitute

$P_1/T_1 = P_2/T_2$
(4.000 atm)/(473 K) = (20.00 atm)/T_2
$T_2 = 2365$ K

The **combined gas law** is the combination of Boyle's and Charles' laws that allows for temperature, pressure, and volume to all be variable.

$P_1V_1/T_1 = P_2V_2/T_2$
where
P_1 = initial pressure
V_1 = initial volume
T_1 = initial temperature (in Kelvin)
P_2 = final pressure
V_2 = final volume
T_2 = final temperature (in Kelvin)

Example 4

60 liters of gas at 350 K and 800 kPa is heated to 700 K and a pressure of 1,600 kPa. What is the final volume?

$P_1V_1/T_1 = P_2V_2/T_2$
(800 kPa)(60 L)/(350 K) = (1600 kPa)(V_2)/(700 K)
V_2 = 60 L

An ideal gas is a gas that obeys the kinetic model in which all particles in a gas are points that move in constant, random, straight-line motion until collisions occur between them. Such a gas does not actually exist and is therefore only theoretical. However, many gases conform well to this behavior at low pressures, and approximations can be made using the ideal gas law that are fairly good estimates.

The **ideal gas law** is the relationship expressed by the formula

$PV = nRT$
where
P = pressure (in kPa)
V = volume (in liters)
n = moles of gas (in moles)
R = ideal gas law constant = 8.3145 L·kPa/mol·K
T = temperature (in Kelvin)

Note that to use this particular constant, the units used in calculations must be as listed in units of the ideal gas law constant. If atmospheres are used then the value would be 0.08206 L·atm/mol·K.

One of the properties of gases that makes this approximation work so well was discovered by Italian scientist Amedeo Avogadro when he determined that the volume of 1 mole of gas was 22.4 liters no matter what kind of gas was used, assuming that the various gases were all under the same temperature and pressure.

Example 5

What volume is occupied by 1.5 moles of gas at 325 K, under 94.0 kPa of pressure?

$PV = nRT$

$(94.0\ \text{kPa})(V) = (1.5\ \text{mol})(8.3145\ \text{L·kPa/mol·K})(325\ \text{K})$

$V = 43\ \text{L}$

Exploration Activities

Solve the following gas law problems.

1. A sample of gas has a volume of 400 cm^3 and is under a pressure of 2.3 atm. If the temperature is held constant and the volume is increased to 600 cm^3, what will the final pressure be?

2. A sample of gas has a volume of 4.5 liters at a temperature of 211 K while at a constant pressure. If the volume is decreased to 2.25 liters, what will the final temperature be?

3. A sample of gas has a pressure of 56 kPa at a temperature of 273 K while the volume is held constant. If the temperature is decreased to 100 K, what will the new pressure be?

4. A sample of gas has a volume of 550 cm^3 at 300 K when the pressure is 16 atm. If the pressure is decreased to 8 atm and the temperature increased to 400 K, what will the new volume be?

5. What is the volume in cm^3 of 3.4 moles of nitrogen gas at 220 kPa and 275°C?

6. A sample of gas has a volume of 16 liters and is under a pressure of 203.4 kPa. If the temperature is held constant and the pressure is increased to 314.2 kPa, what will the final volume be?

7. A sample of gas has a temperature of 211°C and a volume of 300 cm^3 while at a constant pressure. If the temperature is increased to 411°C, what will the new volume be?

8. A sample of gas has a pressure of 6.4 atm at a temperature of 200 K while at a constant volume. If the pressure increases to 8.4 atm, what will the new temperature be?

9. A sample of gas has a volume of 4.4 liters at 300°C and 122 kPa. If the volume is increased to 10.2 liters and the temperature decreased to 250°C, what is the new pressure?

10. How many moles of methane gas are needed to occupy 4.3 liters at a temperature of 400 K and a pressure of 2 atm?

Student Lab: Boyle's Law

As the pressure changes in a balloon, the size of the balloon changes as well. This relationship is something we use every time we inflate a balloon, a tire, or a basketball. There is an exact connection between the pressure and volume, and the goal of this lab is to discover that relationship. One of the key factors is to try and see what happens while the temperature is held constant, because we also know that when gases get hotter they tend to expand.

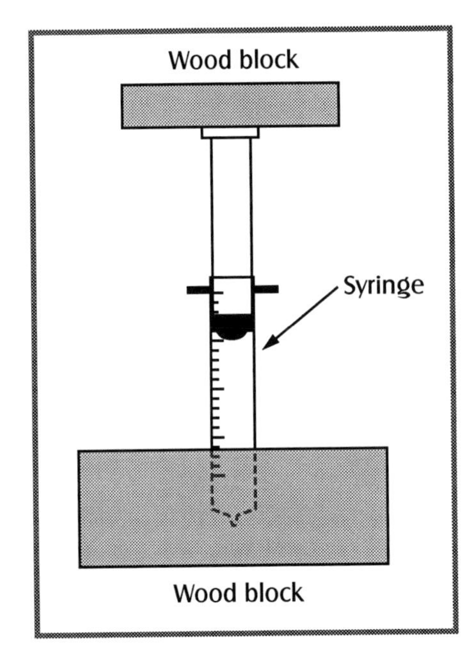

Materials

- Boyle's law apparatus, or approximation of syringe setup as shown to the left
- 60 milliliter or larger plastic syringe with a sealed tip
- platform top for syringe
- kilogram and 500-gram masses

Safety Considerations

The masses used may become unbalanced when being placed on the top platform. Students should set up the lab so that falling masses don't land on their feet. Students should also place a book or thick newspaper under the apparatus to prevent damage to lab counters or desktops.

Procedure

1. Measure the mass of the syringe plunger and wooden platform from the top of the apparatus.
2. Push the plunger back into the syringe, let it come to equilibrium, and record the volume of gas under the plunger.
3. Add a 500-gram mass to the platform and record the volume of gas under the plunger. Repeat this in 500-gram steps up to 3 kilograms and record each volume in a chart. (For example, 0.5 kilogram, 1.0 kilogram, 1.5 kilograms, etc.)

Calculations

Calculate the pressure exerted on the gas for each mass, including just the plunger and platform.

Comprehension Questions

1. What is the effect of an increase in pressure on the volume of the trapped gas?

2. Determine your *PV* products (for each row in your chart—multiply pressure and volume).

3. Plot a graph of volume versus pressure.

4. Using your graph, determine the relationship between pressure and volume.

Answer Key

Matter and Energy

1. Joules
2. The mass should be the same.
3. Food energy → chemical potential energy → mechanical energy

Mass and Weight

1. Answers will vary, but milligrams are a common mass unit used for measuring medications.
2. Both contain the same amount of matter; it is their mass that is different.
3. 500 kg on the Moon

States of Matter

1. Liquid
2. Solid
3. Answers will vary, but shortening can be seen in all three forms when melted for deep frying if it is heated enough to boil.
4. A crystal is a material with a repeating geometric pattern made of atoms, molecules, or ions.
5. By distillation

Student Lab: Change of Phase

2. The lines should be horizontal around the melting and boiling point, indicating that temperature does not change while the state is changing.
5. Purity of ice sample, barometric pressure, etc.

Chemical and Physical Properties

1. a. Physical
   b. Chemical
   c. Physical
   d. Chemical
   e. Chemical
2. Answers will vary, but should be both chemical and physical.
3. Physical properties can be tested without damaging the sample.

Chemical and Physical Changes

1. Breaking, tearing, crushing, bending, ripping
2. Burning, corroding, reacting with acid, reacting with base, combining with another substance
3. Chopping wood and burning it

Categories of Material

1. mixture
2. mixture
3. substance, compound
4. mixture
5. mixture
6. substance, element

Units of Measurement

1. Answers will vary—miles/hour, miles/gallon, pounds/square inch, etc.

2. So that people of a wide variety of languages and backgrounds can have a system that allows them to communicate clearly
3. a. 2
   b. 3
   c. 4
   d. 2
   e. 5
   f. 3

Density and Specific Gravity

1. 1.2 g/cm^3
2. 8.96
3. Yes

Heat and Temperature

1. Endothermic
2. Answers will vary—stove to pan, water in pipes to radiator surface, etc.
3. Answers will vary—hair dryer, fan, convection oven, forced hot air furnace.

Student Lab: Physical Properties of Compounds

1. State at room temp, conductivity, etc.
2. State at room temp, color, crystalline nature
3. Answers will vary.
4. Taste
5. According to the lab results, salt and sugar are polar, while cooking oil is nonpolar.

Student Lab: Chemical Properties of Substances

1. Bubbles forming on the surface
2. Melting
3. It had formed a compound.

Student Lab: Determining Density

1. No, there is no simple formula for the area of an irregularly shaped object.
2. Answers will vary.
3. Mass it, find volume using $\pi r^2 h$, or by water displacement, then m/V.

Atoms and Molecules

1. We know there are smaller particles; he did not have the technology to prove or disprove his theory.
2. By mass
3. An atom of an element that has a different number of neutrons

Chemical Formulas

1. a. Hydrogen, sulfur, oxygen
   b. Barium, chlorine
   c. Carbon, hydrogen
2. a. Li_2O
   b. CCl_4
   c. Ag_3PO_4

Ions

1. Na^{1+}
2. Zero
3. Li_2O

Atomic Mass

1. 6.02×10^{23}
2. 180 u

3. $(30.97\ u)(1.6605 \times 10^{-24}\ g/u) = 5.143 \times 10^{-23}\ g$, 30.97 u

Skill Builder: Calculating Formula Mass

1. $NaCl = 58.5$ u
2. $H_2O = 18.0$ u
3. $KCl = 74.6$ u
4. $MgO_2 = 56.3$ u
5. $BaCO_3 = 197.3$ u
6. $Fe_2O_3 = 159.3$ u
7. $P_4O_{10} = 284$ u
8. $CH_4 = 16.0$ u
9. $C_5H_{12} = 72.0$ u
10. $C_6H_5OH = 94.0$ u
11. $NH_4Cl = 53.5$ u
12. $AgNO_3 = 169.6$ u
13. $CaCO_3 = 100.1$ u
14. $BF_3 = 67.8$ u
15. $Ca(OH)_2 = 74.1$ u
16. $HCl = 36.5$ u
17. $PbSO_4 = 303.3$ u
18. $MgCO_3 = 84.3$ u
19. $I_2 = 253.8$ u
20. $FeS_2 = 120.0$ u
21. $KOH = 56.1$ u
22. $KClO_4 = 138.6$ u
23. $SnO_2 = 150.7$ u
24. $C_6H_6 = 78.0$ u
25. $Hg_2Cl_2 = 271.6$ u
26. $HNO_3 = 63.0$ u
27. $PbBr_2 = 367.0$ u
28. $CaO = 56.1$ u
29. $H_2SO_4 = 98.1$ u
30. $Al(OH)_3 = 78.0$ u

Student Lab: Empirical Formula

1. a. MgO
   b. Mg_2O_3
   c. MgO_2
   d. Mg_2O_5
2. a.

Solutions

1. An indicator
2. 2.5 moles
3. 0.75 molar (0.75 M)

The Periodic Table

1. By atomic mass
2. a. metal
   b. metal
   c. nonmetal
   d. metalloid
3. Metals usually lose electrons.

Types of Reactions

(sr = single replacement, dr = double replacement, s = synthesis, d = decomposition, c = combustion)

1. dr
2. s
3. c
4. dr
5. sr
6. d
7. s
8. sr
9. dr
10. d
11. sr
12. c
13. s
14. d
15. dr
16. d
17. sr
18. $PbCrO_4(s)$
19. $Ba(IO_3)_2(s)$

Acid-Base Reactions

1. $H_2SO_4 + 2NaOH \rightarrow Na_2SO_4 + 2H_2O$

H_2SO_4	$2NaOH$	Na_2SO_4	$2H_2O$
Acid	Base	Salt	Water

2. a. 4
   b. 8
   c. 12
3. a. 1.0×10^{-7} M
   b. 1.0×10^{-9} M
   c. 1.0×10^{-6} M
   d. 1.0×10^{-2} M

Student Lab: Chemical Reaction Types

Single-replacement reactions

Reaction	Observation
1. $Zn(s) + CuSO_4(aq) \rightarrow Cu(s) + ZnSO_4(aq)$	Copper solid forms on zinc.
2. $Zn(s) + AgNO_3(aq) \rightarrow Ag(s) + ZnSO_4(aq)$	Shiny silver crystals grow on zinc solid.
3. $Zn(s) + Pb(NO_3)_2(aq) \rightarrow Pb(s) + Zn(NO_3)_2(aq)$	Dull gray lead forms on pieces of zinc.
4. $Zn(s) + 2HCl(aq) \rightarrow ZnCl_2(aq) + H_2(g)$	Zinc reacts with hydrochloric acid and hydrogen gas is released.
5. $2Al(s) + 3CuSO_4(aq) \rightarrow 3Cu(s) + Al_2(SO_4)_3(aq)$	Copper forms on aluminum.
6. $Cu(s) + 2AgNO_3(aq) \rightarrow 2Ag(s) + Cu(NO_3)_2(aq)$	Shiny silver crystals grow on copper solid.
7. $Cu(s) + Pb(NO_3)_2(aq) \rightarrow$ No reaction	Copper will not replace lead.
8. $2Al(s) + 6HCl(aq) \rightarrow 2AlCl_3(aq) + 3H_2(g)$	Aluminum reacts with hydrochloric acid and hydrogen gas is released.

Double-replacement reactions

Reaction	Observation
1. $NaCl(aq) + CuSO_4(aq) \rightarrow$ No reaction	All products are soluble.
2. $2NaOH(aq) + CuSO_4(aq) \rightarrow Na_2SO_4(aq) + Cu(OH)_2(s)$	Gelatinous blue copper (II) hydroxide precipitates.
3. $2NaOH(aq) + Pb(NO_3)_2(aq) \rightarrow 2NaNO_3(aq) + Pb(OH)_2(s)$	Dull gray lead (II) hydroxide precipitates.
4. $NaCl(aq) + AgNO_3(aq) \rightarrow NaNO_3(aq) + AgCl(s)$	Shiny silver chloride precipitates.
5. $2NaCl(aq) + Pb(NO_3)_2(aq) \rightarrow 2NaNO_3(aq) + PbCl_2(s)$	Lead (II) chloride precipitates.
6. $NaOH(aq) + AgNO_3(aq) \rightarrow NaNO_3(aq) + AgOH(s)$	Gray silver hydroxide precipitates.

Oxidation Numbers

1. They are the same number.
2. Al^{3+} and O^{2-}
3. $S^{4+} + 3O^{2-} \rightarrow SO_3^{2-}$ or $4 + 3(-2) = -2$

Oxidation–Reduction Reactions

1. Oxidation is the loss of electrons and reduction is the gaining of electrons. The oxidation number of an atom that undergoes oxidation is increased in value.
2. a. $CaBr_2 + Pb(NO_3)_2 \rightarrow Ca(NO_3)_2 + PbBr_2$

   No oxidation or reduction takes place in this reaction.

   b. $SnCl_2 + 2FeCl_3 \rightarrow SnCl_4 + 2FeCl_2$

   Tin changes from +2 to +4, and iron from +3 to +2.

   c. $4HCl + MnO_2 \rightarrow 2H_2O + Cl_2 + MnCl_2$

   Mn changes from +4 to +2, and chlorine from −1 to 0.

Balancing Chemical Equations

1. To accurately predict the amount of reactants needed and the amount of products produced.
2. $CaSO_4 + Na_2CO_3 \rightarrow Na_2SO_4 + CaCO_3$ (balanced as written)
3. $2ZnS + 3O_2 \rightarrow 2ZnO + 2SO_2$

Skill Builder: Balancing Chemical Equations

1. $2Na + 2H_2O \rightarrow 2NaOH + H_2$
2. $CaCO_3 + 2HCl \rightarrow CO_2 + CaCl_2 + H_2O$
3. $N_2 + 3H_2 \rightarrow 2NH_3$
4. $Ca(OH)_2 + H_2SO_4 \rightarrow CaSO_4 + 2H_2O$
5. $2C_2H_2 + 5O_2 \rightarrow 4CO_2 + 2H_2O$
6. $2NaCl \rightarrow Cl_2 + 2Na$
7. $3Ti + 2N_2 \rightarrow Ti_3N_4$
8. $2Al + 3ZnCl_2 \rightarrow 2AlCl_3 + 3Zn$
9. $4Al + 3O_2 \rightarrow 2Al_2O_3$
10. $4Fe + 3O_2 \rightarrow 2Fe_2O_3$
11. $2CuO \rightarrow 2Cu + O_2$
12. $2NO \rightarrow N_2 + O_2$

13. $NaCl + Li \rightarrow LiCl + Na$

14. $2NaI + Br_2 \rightarrow 2NaBr + I_2$

15. $CaCO_3 + 2NH_4Cl \rightarrow (NH_4)_2CO_3 + CaCl_2$

16. $2KOH + H_2SO_4 \rightarrow K_2SO_4 + 2H_2O$

17. $2Al + 3H_2SO_4 \rightarrow Al_2(SO_4)_3 + 3H_2$

18. $3Ca + 2H_3PO_4 \rightarrow Ca_3(PO_4)_2 + 3H_2$

19. $Mg + 2HNO_3 \rightarrow Mg(NO_3)_2 + H_2$

20. $C_{25}H_{52} + 38O_2 \rightarrow 25CO_2 + 26H_2O$

21. $4Zn + 10HNO_3 \rightarrow 4Zn(NO_3)_2 + NH_4NO_3 + 3H_2O$

22. $4NH_3 + 6NO \rightarrow 5N_2 + 6H_2O$

23. $5C + 2SO_2 \rightarrow CS_2 + 4CO$

24. $Sb_2S_3 + 3Fe \rightarrow 2Sb + 3FeS$

25. $3CaCO_3 + 2H_3PO_4 \rightarrow Ca_3(PO_4)_2 + 3CO_2 + 3H_2O$

Structure of the Atom

1. a. Leptons—a group of particles that includes electrons, muons, and tau particles

   b. Atomic number—the number of protons in the nucleus of an atom

   c. Photoelectric effect—the ejection of photons from the surface of a material when it is bombarded with energy

2. By shooting alpha particles at a sheet of gold foil a few atoms thick

3. The proton and neutron have very similar masses, and the electron is about 1/2000th the mass of either.

4. 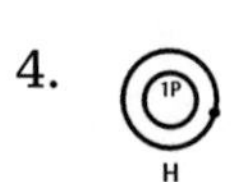

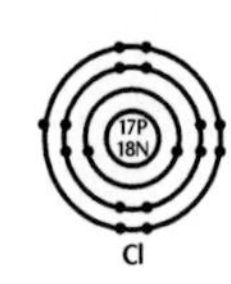

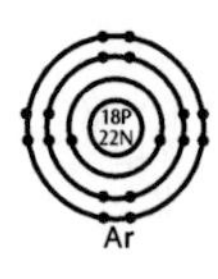

5. Red light

Student Lab: Atomic Structure and the Flame Test

Metal	Flame color
Barium	Light green
Calcium	Brick red
Copper	Blue/green
Lithium	Dark red/orange
Potassium	Lilac
Sodium	Bright orange

1. Answers will vary depending on which samples are chosen as unknowns.

2. There is no way to tell the subtle differences between colors with just the naked eye as the only measuring tool.

3. Physical and chemical properties could be measured for each substance; also, a spectrometer could be used to add numerical values to the colors measured.

Electron Configurations

1. a. Lithium = $1s^2\ 2s^1$
   b. Magnesium = $1s^2\ 2s^2\ 2p^6\ 3s^2$
   c. Argon = $1s^2\ 2s^2\ 2p^6\ 3s^2\ 3p^6$
   d. Iodine = $1s^2\ 2s^2\ 2p^6\ 3s^2\ 3p^6\ 4s^2\ 3d^{10}\ 4p^6\ 5s^2\ 4d^{10}\ 5p^5$
   e. Rubidium = $1s^2\ 2s^2\ 2p^6\ 3s^2\ 3p^6\ 4s^2\ 3d^{10}\ 4p^6\ 5s^1$
2. a. Lithium = 1
   b. Magnesium = 2
   c. Argon = 8
   d. Iodine = 7
   e. Rubidium = 1
3. a. Lithium = Li
   b. Magnesium = Mg
   c. Argon = Ar
   d. Iodine = I
   e. Rubidium = Rb
4. a. NaCl = linear
   b. BeH_2 = linear
   c. SO_2 = bent
   d. SO_3 = trigonal planar
   e. H_2O = bent
   f. NH_3 = pyramidal
   g. CCl_4 = tetrahedral

Student Lab: Chemical Bonding

1. For multiple nuts and bolts it solves the problem of any one nut or bolt being off a little from the others in mass.
2. It takes into account the idea that not all atoms of an element have the same mass, because some isotopes have more mass than others.
3. Will depend on the average mass of the nuts and bolts used in the lab

Gas Laws

1. 1.53 atm
2. 105.5 K
3. 20.5 kPa
4. 1467 cm^3
5. 70.4 liters = 70,400 cm^3
6. 10.4 liters
7. 424 cm^3
8. 262.5 K
9. 48.0 kPa
10. 0.262 mole

Student Lab: Boyle's Law

1. The volume of the gas decreases.
4. Students should find that $PV = K$.

Rubric: Assessing Laboratory Reports

This book contains several student laboratory assignments for which you will produce a written report. Lab reports are important because they are a written recipe for another scientist to replicate your findings. Information should include:

- *Purpose:* Why is this lab being performed? What is the objective of the lab?
- *Hypothesis:* Given the initial level of knowledge, what do you expect to find out at the end?
- *Materials list:* A well-organized materials list makes it easier for anyone trying to replicate your results to understand what you did.
- *Procedure:* Even though a procedure is suggested in the lab write-ups, you should include the procedure you actually followed.
- *Data:* What actually took place in the lab?
- *Conclusion:* What were the results? Did your hypothesis match the data? If something went wrong, what do you think happened?

Below is a rubric your teacher may use to assess your lab reports.

	1	2	3	4
Understanding of concept	**poor**	**adequate**	**good**	**outstanding**
Methodology	**poor**	**adequate**	**good**	**outstanding**
Organization of experiment	**poor**	**adequate**	**good**	**outstanding**
Organization of report	**poor**	**adequate**	**good**	**outstanding**

Rubric: Assessing Essays

In addition to lab reports, you will also be assigned several essays. You should take this opportunity to do outside research on the subject matter to supplement the material provided in this book.

Below is a rubric your teacher may use to assess your essays.

	1	2	3	4
Quality of research	poor	adequate	good	outstanding
Organization of material	poor	adequate	good	outstanding
Presentation of material	poor	adequate	good	outstanding
Spelling, grammar, and style	poor	adequate	good	outstanding

Ion Chart

IONS $^{1+}$

Cesium, Cs^{1+}

Copper (I), Cu^{1+}

Hydrogen, H^{1+}

Indium, In^{1+}

Lithium, Li^{1+}

Potassium, K^{1+}

Rubidium, Rb^{1+}

Silver, Ag^{1+}

Sodium, Na^{1+}

Thallium, Tl^{1+}

IONS $^{2+}$

Barium, Ba^{2+}

Beryllium, Be^{2+}

Cadmium, Cd^{2+}

Calcium, Ca^{2+}

Chromium(II), Cr^{2+}

Cobalt(II), Co^{2+}

Copper(II), Cu^{2+}

Iridium(II), Ir^{2+}

Iron(II), Fe^{2+}

Lead(II), Pb^{2+}

Magnesium, Mg^{2+}

Manganese(II), Mn^{2+}

Mercury(II), Hg^{2+}

Nickel(II), Ni^{2+}

Platinum(II), Pt^{2+}

Strontium(II), Sr^{2+}

Tin(II), Sn^{2+}

Titanium(II), Ti^{2+}

Tungsten(II), W^{2+}

Vanadium(II), V^{2+}

Zinc, Zn^{2+}

Zirconium(II), Zr^{2+}

IONS $^{3+}$

Aluminum, Al^{3+}

Antimony(III), Sb^{3+}

Bismuth(III), Bi^{3+}

Boron, B^{3+}

Cerium(III), Ce^{3+}

Cobalt(III), Co^{3+}

Chromium(III), Cr^{3+}

Gallium(III), Ga^{3+}

Indium(III), In^{3+}

Iridium(III), Ir^{3+}

Iron(III), Fe^{3+}

Manganese(III), Mn^{3+}

Phosphorus(III), P^{3+}

Rhodium(III), Rh^{3+}

Thallium(III), Tl^{3+}

Titanium(III), Ti^{3+}

Uranium(III), U^{3+}

Vanadium(III), V^{3+}

IONS $^{4+}$

Cerium(IV), Ce^{4+}

Germanium(IV), Ge^{4+}

Iridium(IV), Ir^{4+}

Lead(IV), Pb^{4+}

Platinum(IV), Pt^{4+}

Thorium(IV), Th^{4+}

Titanium(IV), Ti^{4+}

Tin(IV), Sn^{4+}

Tungsten(IV), W^{4+}

Uranium(IV), U^{4+}

Vanadium(IV), V^{4+}

Zirconium(IV), Zr^{4+}

IONS $^{5+}$

Antimony(V), Sb^{5+}

Bismuth(V), Bi^{5+}

Phosphorus(V), P^{5+}

Tungsten(V), W^{5+}

Uranium(V), U^{5+}

Vanadium(V), V^{5+}

IONS $^{1-}$

Bromide, Br^{1-}

Chloride, Cl^{1-}

Fluoride, F^{1-}

Hydride, H^{1-}

Iodide, I^{1-}

IONS $^{2-}$

Oxide, O^{2-}

Selenide, Se^{2-}

Sulfide, S^{2-}

IONS $^{3-}$

Nitride, N^{3-}

Phosphide, P^{3-}

IONS $^{4-}$

Carbide, C^{4-}

POLYATOMIC IONS $^{1+}$

Ammonium, $NH_4^{\ 1+}$

Hydronium, H_3O^{1+}

POLYATOMIC IONS $^{2+}$

Mercury(I), $Hg_2^{\ 2+}$

POLYATOMIC IONS $^{1-}$

Acetate, $C_2H_3O_2^{\ 1-}$

Azide, $N_3^{\ 1-}$

Benzoate, C_6HCOO^{1-}

Bromate, $BrO_3^{\ 1-}$

Chlorate, $ClO_3^{\ 1-}$

Chlorite, $ClO_2^{\ 1-}$

Cyanide, CN^{1-}

Formate, $HCOO^{1-}$

Hydrogen carbonate, or Bicarbonate, $HCO_3^{\ 1-}$

Hydrogen sulfate, $HSO_4^{\ 1-}$

Hydrogen sulfide, HS^{1-}

Hydrogen sulfite, $HSO_3^{\ 1-}$

Hydroxide, OH^{1-}

Hypochlorite, ClO^{1-}

Iodate, $IO_3^{\ 1-}$

Nitrate, $NO_3^{\ 1-}$

Nitrite, $NO_2^{\ 1-}$

Perchlorate, $ClO_4^{\ 1-}$

Periodate, $IO_4^{\ 1-}$

Permanganate, $MnO_4^{\ 1-}$

POLYATOMIC IONS $^{2-}$

Carbonate, $CO_3^{\ 2-}$

Chromate, $CrO_4^{\ 2-}$

Dichromate, $Cr_2O_7^{\ 2-}$

Hexafluorosilicate, $SiF_6^{\ 2-}$

Hydrogen phosphate, $HPO_4^{\ 2-}$

Molybdate, $MoO_4^{\ 2-}$

Oxalate, $C_2O_4^{\ 2-}$

Peroxide, $O_2^{\ 2-}$

Selenate, $SeO_4^{\ 2-}$

Silicate, $SiO_3^{\ 2-}$

Sulfate, $SO_4^{\ 2-}$

Sulfite, $SO_3^{\ 2-}$

Tartrate, $C_4H_4O_6^{\ 2-}$

Thiosulfate, $S_2O_3^{\ 2-}$

POLYATOMIC IONS $^{3-}$

Arsenate, $AsO_4^{\ 3-}$

Borate, $BO_3^{\ 3-}$

Phosphate, $PO_4^{\ 3-}$

Phosphite, $PO_3^{\ 3-}$

Summary of VSEPR Geometries

Number of sigma bonding and nonbonding electron pairs about central atom	Number of nonbonding pairs	Examples	Shape
2	0	N_3^-, CO_2	linear
3	0	BCl_3, CO_3^{2-}	trigonal planar
3	1	NO_2^-	bent
4	0	CH_4, SO_4^{2-}	tetrahedral
4	1	NH_3, PCl_3	pyramidal
4	2	H_2O	bent
5	0	PCl_5	trigonal biparamidal
5	1	SF_4	see-saw
5	2	BrF_3	T-shaped
5	3	I_3^-	linear
6	0	SiF_6^{2-}	octahedral
6	1	BrF_5	square pyramidal
6	2	ICl_4^-, XeF_4	square planar

Scientific Suppliers

Here is a short list of reliable scientific suppliers who will sell to home and school markets.

Edmund Scientifics
60 Pearce Avenue
Tonawanda, NY 14150-6711
800-728-6999 (telephone)
800-828-3299 (fax)
www.scientifics.com
Biology, earth science, chemicals (after providing forms)

Fisher Science Education
2000 Park Lane
Pittsburgh, PA 15275
800-955-1177
www.fisheredu.com or www.fishersci.com
Biology, earth science, physics, labware

Educational Innovations, Inc.
362 Main Avenue
Norwalk, CT 06851
888-912-7474 (telephone)
203-229-0740 (fax)
www.teachersouce.com
Biology, earth science, physics, labware

Frey Scientific
100 Paragon Parkway
Mansfield, OH 44903
800-225-3739
www.freyscientific.com
Biology, earth science, physics, chemicals, labware

Time Line of Chemistry

c. 700 B.C.E.—Greeks discovered electric attraction produced by rubbing amber.

c. 600 B.C.E.—Thales proposed that nature should be understood by replacing myth with logic and that all matter is made of water.

c. 500 B.C.E.—Anaximenes introduced the ideas of condensation and rarefaction.

c. 450 B.C.E.—Anaxagoras proposed the first clearly materialist philosophy—the universe is made entirely of matter in motion.

c. 370 B.C.E.—Leucippus and Democritus proposed that matter is made of small, indestructible particles.

1232—Rockets are invented in China to defend city of Kaifeng against Mongol invaders.

1600—Gilbert discovered that electricity occurs in things other than amber and wrote a book on magnetism.

1620—Bacon published *Novum Organum* (scientific method and inductive reasoning).

1652—Pascal discovered laws of fluid pressure.

1738—Bernoulli proposed laws of fluid mechanics.

1777—Lavoisier proposed idea of chemical compounds made of elements.

1785—Coulomb confirmed the inverse square law for electric force.

1787—Berthollet proposed system of chemical nomenclature.

1800—Volta invented the battery.

1800—Ampere discovered properties of magnetic field produced by electric current.

1803—Dalton composed the law of definite proportions in chemistry.

1808—Dalton published a periodic table based on atomic weights.

1811—Avogadro introduced the concept of the mole.

1825—discovery of Ampere's force law

1827—Brown discovered Brownian motion.

1869—Mendel used a periodic table of known elements to correctly predict the properties of then undiscovered elements.

1873—Maxwell published *Treatise on Electricity and Magnetism.*

1883—Wroblewski and Olszewski first produced liquid oxygen.

1885—Hertz discovered the photoelectric effect.

1885—Balmer discovered spectral lines of hydrogen.

1898—Curie and Curie announced their discovery of radium and polonium.

1900—Planck proposed that energy can only be absorbed or emitted by matter in discrete amounts (quanta).

1905—Einstein published papers on Brownian motion, the photoelectric effect, and the special theory of relativity.

1911—Rutherford discovered that the positive charge in an atom is concentrated in a small nucleus and proposed a planetary model of the atom.

1913—Bohr published his model of the atom, based on energy states described by one quantum number.

1916—Lewis proposed the idea of covalent bonds.

1924—De Broglie proposed that all matter has wave properties.

1925—Pauli proposed the Exclusion Principle (no two electrons in an atom can have the same set of quantum numbers).

1926—Schrodinger developed the wave equation.

1926—Born proposed the statistical interpretation of the wave equation.

1927—Heisenberg proposed the Uncertainty Principle (we cannot simultaneously determine the position and momentum of a subatomic particle).

1927—Experiment by Davisson and Germer, and simultaneous experiment by G. P. Thompson, proved the wave behavior of electrons.

1931—Anderson discovered the positron.

1932—Chadwick discovered the neutron.

1937—discovery of the muon

1938—Hahn, Strassmann, Meitner, and Frisch discovered nuclear fission.

1947—W. F. Libby invented radiocarbon dating.

1947—discovery of the pion (predicted by Yukawa in 1935)

1951—Franklin discovered nucleic acids (RNA and DNA), helical shape.

1953—Miller produced amino acids from inorganic compounds and sparks.

1956—discovery of the neutrino (predicted by Pauli in 1930)

1963—Gell-Mann proposed protons and neutrons are made of smaller particles (quarks).

1995—discovery of the top quark at Fermilab

acid: (1) a substance that can be a proton donor (2) a substance that can act as an electron acceptor (3) a substance that produces hydronium ions when dissolved in water

adhesion: the molecular attraction between the surfaces of two unlike bodies in contact

amorphous solid: a solid in which the atoms or molecules that make up the solid are in a disorganized arrangement

anion: a negative ion

atom: the smallest particle of an element that contains all the properties of that element

atomic mass: the mass of an atom as measured in atomic mass units

atomic mass unit: the unit of mass equal to $\frac{1}{12}$ the mass of a carbon-12 atom

atomic number: the number of protons in the nucleus of an atom

Avogadro's number: the number of atoms in one mole of a substance, equal to 6.02×10^{23}

base: (1) a substance that can be a proton acceptor (2) a substance that can act as an electron donor (3) a substance that produces hydroxyl ions when dissolved in water

boiling point: the temperature at which the vapor pressure of a liquid is equal to the atmospheric pressure at the surface of the liquid

Boyle's law: the statement that the volume of a gas is inversely proportional to its pressure at a constant temperature

capillary action: the tendency of a liquid surface to follow a surface it is in contact with because of adhesion and surface tension

cation: a positive ion

charge: a positive or negative change in a particle away from being electrically neutral

Charles' law: the statement that the volume of a gas is directly proportional to its temperature at a constant pressure

chemical change: as the result of a chemical reaction, the change of the properties and composition of a substance

chemical equation: a shorthand statement that uses chemical formulas to describe a chemical reaction

chemical formula: a shorthand method using numbers and chemical symbols to describe the makeup of one molecule of a substance

chemical property: a characteristic of a substance that is observed as the substance undergoes a chemical change

combination reaction (synthesis): a reaction in which two or more elements combine to form a single substance

combined gas law: the combination of Boyle's and Charles' laws that allows for temperature, pressure, and volume to all be variable

combustion reaction (hydrocarbon combustion): a reaction in which a hydrocarbon is burned and consumed in the presence of oxygen

compound: a substance made of atoms of two or more elements combined in a definite proportion

concentration: the amount of substance in a given volume

conduction: the movement of heat or some other form of energy through a material as the heat or energy is passed from particle to particle

convection: the transfer of heat from one body to another by a fluid such as air or water

corrosion: a reaction like rusting that results in the unwanted comsumption of one or more of the substances, usually a metal

covalent bond: a bond formed by the sharing of electrons, usually a pair, between two atoms

crystal: a solid in which the constituent particles (atoms, ions, molecules) form an orderly, repeating pattern

crystal lattice: the repeating pattern of atoms, ions, or molecules in a crystalline solid

decomposition reaction (analysis): a reaction in which a single substance is separated into two separate substances

density: a measure of the amount of mass in a given volume

derived units: the combination of base units into a single unit, like meters (m) and seconds (s) into meters per second (m/s)

dilution: the process of reducing the concentration of a solution, usually by adding more solvent

dipole: a molecule in which there are positive and negative locations despite the fact that the molecule as a whole is electrically neutral

distillation: the process of vaporizing and condensing a substance or mixture to purify or separate its various components

double-replacement reaction: a reaction in which two ionic compounds exchange ions to create two new ionic compounds

electrolytes: substances whose aqueous solutions will conduct electricity

electromagnetic radiation: an oscillating wave having both electric and magnetic components; includes parts of the electromagnetic spectrum such as light and X rays

electron: one of the three major subatomic particles, it has a negative charge, is found outside of the nucleus of the atom, and is symbolized as e^-.

electron configuration: the arrangement of electrons in the various orbitals of an atom

element: a substance that cannot be separated into simpler substances by ordinary chemical means

empirical formula: the simplest ratio in which atoms can combine to form a compound

endothermic: a chemical reaction or process that absorbs heat

energy: the ability of a system to do work; it is conserved throughout a chemical reaction

evaporation: when particles in the liquid phase escape the liquid and enter the gas phase, usually at a temperature other than the boiling point

exothermic: a chemical reaction or process that releases heat

gas: the fluid state of matter in which the molecules are widely spread apart; a gas has neither a definite shape nor volume.

Gay-Lussac's law: the statement that the pressure of a gas varies directly with the Kelvin temperature at a constant volume

heat: a form of energy that can be transferred between two bodies of differing temperature

hydrogen bonds: very weak bonds that can be formed between a polar molecule with hydrogen and another molecule that is polar

ideal gas law: the relationship represented by the equation $PV = nRT$

inertia: the tendency of matter to resist any change in its motion

inorganic compound: typically a compound that is not organic

ion: an atom or a group of atoms that has acquired a charge due to a loss or gain of electrons

ionic bond: a bond formed by the attraction between a cation and an anion

isotopes: atoms of an element that have different atomic masses because they have different numbers of neutrons

liquid: the fluid state of matter in which particles are close together; a liquid has a definite volume and an indefinite shape.

mass: a measure of the amount of matter in an object

matter: any material that has mass and occupies space

melting point: the temperature at which a substance changes from a solid to a liquid

metal: an element that loses electrons easily in a chemical change and has the properties of high luster, electrical and thermal conductivity, malleability, and ductility

metalloid (semimetal): an element that has both metallic and nonmetallic properties

mixture: a combination of two or more substances that do not chemically combine and can be separated by physical means

molar mass: the mass of 1 mole of particles of a substance

molarity: the concentration of a solution given in moles of solute per liter of solution

mole: the SI unit of amount of matter, often given as 6.02×10^{23} particles (atoms, molecules, ions) of a substance

molecular mass: the mass of one molecule of a substance, given in atomic mass units

molecule: the smallest particle of a compound that maintains all the properties of that compound

neutralization reaction: the reaction of an acid with a base that results in salt and water as the products

neutron: one of the three major subatomic particles, it is electrically neutral and is found in the nucleus of the atom; it is similar in mass to the proton.

newton: the SI unit of force

nonmetal: an element that easily gains electrons during a chemical reaction and whose properties contrast with those of the metals

organic compound: a compound containing carbon, usually with hydrogen

oxidation: a reaction in which a particle, such as an ion, atom, or molecule, loses an electron

oxidation number: the effective charge of an atom when it is chemically combined in a compound

oxidation-reduction reaction: a reaction in which both an oxidation reaction and a reduction reaction take place

periodicity: the idea that there are recurring patterns in the properties of elements when the elements are placed in order according to their atomic numbers

periodic table: a chart showing the elements in order by increasing atomic number and grouped by their similar qualities

photoelectric effect: the result of bombarding the surface of some metals with electromagnetic radiation where the metal emits electrons

physical change: a change in a material in which the physical properties of the substance, but not the chemical properties, are changed

physical property: a characteristic of a substance that can be observed without production of new chemical materials

plasma: the fourth state of matter often characterized as an ionized gas

polyatomic ion: a group of two or more elements bonded together and having a collective charge

precipitation: the process in which an insoluble solid is formed in a solution and settles out

pressure: the amount of force exerted on a given area

products: the substances formed in a chemical reaction

proton: one of the three major subatomic particles, it has a positive charge, is found in the nucleus, and has a mass similar to that of the neutron.

radiation: the emissions of energy such as light, X rays, radio waves, alpha particles, etc.

reactants: the substances that combine in a reaction to form products

reduction: a reaction in which a particle, such as an ion, atom, or molecule, gains an electron

semimetal: see metalloid

SI (Système International d'Unités): the International System of Units, a system of basic units for measurement

significant figures: the digits in a measurement that are certain, including one digit in the position of least value that is uncertain

single-replacement reaction: a reaction in which an element replaces a less chemically active element in a compound, and the replaced element is set free

solid: the state of matter that is rigid; a solid has a fixed shape and volume.

solute: that part of a solution that is the dissolved substance

solution: a homogenous mixture of molecular substances

solvent: that part of a solution in which the solute is dissolved

specific gravity: the ratio of the mass of a given volume of any substance to a mass of an equal volume of water

state of matter: the physical condition of a substance, such as solid, liquid, or gas

sublimation: the change of a solid to a vapor without passing through the liquid phase

substance: a pure sample of matter, either an element or a compound

surface tension: the tendency of molecules at the surface of a liquid to be pulled inward

temperature: the measure of the average kinetic energy in the particles of a sample of matter

titration: the process by which the concentration of a solution is determined by reacting it with a solution of known concentration

valence electrons: the electrons in the outermost energy level of an atom, also called the valence shell

valence shell electron pair repulsion (VSEPR): a method for predicting the shape of molecules based on the repulsion of electrons within a molecule

viscosity: the resistance of a fluid to flow

weight: the gravitational force on a sample of matter

Share Your Bright Ideas with Us!

We want to hear from you! Your valuable comments and suggestions will help us meet your current and future classroom needs.

Your name____Date_______________

School name____

School address____

City ____________________State _____Zip_________Phone number (________)____________

Grade level taught________Subject area(s) taught____________________Average class size______

Where did you purchase this publication?_______________________________________

Was your salesperson knowledgeable about this product? Yes_____ No_____

What monies were used to purchase this product?

___School supplemental budget ___Federal/state funding ___Personal

Please "grade" this Walch publication according to the following criteria:

Quality of service you received when purchasing	A	B	C	D	F
Ease of use	A	B	C	D	F
Quality of content	A	B	C	D	F
Page layout	A	B	C	D	F
Organization of material	A	B	C	D	F
Suitability for grade level	A	B	C	D	F
Instructional value	A	B	C	D	F

COMMENTS:____

____

What specific supplemental materials would help you meet your current—or future—instructional needs?

____

Have you used other Walch publications? If so, which ones?______________________________

May we use your comments in upcoming communications? ___Yes ___No

Please **FAX** this completed form to **207-772-3105**, or mail it to:

Product Development, J. Weston Walch, Publisher, P. O. Box 658, Portland, ME 04104-0658

We will send you a **FREE GIFT** as our way of thanking you for your feedback. **THANK YOU!**